DSP for
MATLAB™ and LabVIEW™
Volume II: Discrete Frequency Transforms

SYNTHESIS LECTURES ON SIGNAL PROCESSING

Editor
José Moura, Carnegie Mellon University

DSP for MATLAB™ and LabVIEW™ Volume II: Discrete Frequency Transforms
Forester W. Isen
2008

DSP for MATLAB™ and LabVIEW™
Volume I: Fundamentals of Discrete Signal Processing
Forester W. Isen
2008

The Theory of Linear Prediction
P. P. Vaidyanathan
2007

Nonlinear Source Separation
Luis B. Almeida
2006

Spectral Analysis of Signals
Yanwei Wang, Jian Li, and Petre Stoica
2006

DSP for MATLAB™ and LabVIEW™ Volume II: Discrete Frequency
Transforms Forester W. Isen

ISBN: 978-3-031-01401-7   paperback
ISBN: 978-3-031-02529-7   ebook

DOI 10.1007/978-3-031-02529-7

A Publication in the Springer series
SYNTHESIS LECTURES ON SIGNAL PROCESSING

Lecture #5
Series Editor: José Moura, Carnegie Mellon University

Series ISSN
Synthesis Lectures on Signal Processing
Print 1932-1236   Electronic 1932-1694

DSP for
MATLAB™ and LabVIEW™
Volume II: Discrete Frequency Transforms

Forester W. Isen

SYNTHESIS LECTURES ON SIGNAL PROCESSING #5

ABSTRACT

This book is Volume II of the series *DSP for MATLAB*™ *and LabVIEW*™. This volume provides detailed coverage of discrete frequency transforms, including a brief overview of common frequency transforms, both discrete and continuous, followed by detailed treatments of the Discrete Time Fourier Transform (DTFT), the z-Transform (including definition and properties, the inverse z-transform, frequency response via z-transform, and alternate filter realization topologies including Direct Form, Direct Form Transposed, Cascade Form, Parallel Form, and Lattice Form), and the Discrete Fourier Transform (DFT) (including Discrete Fourier Series, the DFT-IDFT pair, DFT of common signals, bin width, sampling duration, and sample rate, the FFT, the Goertzel Algorithm, Linear, Periodic, and Circular convolution, DFT Leakage, and computation of the Inverse DFT). The entire series consists of four volumes that collectively cover basic digital signal processing in a practical and accessible manner, but which nonetheless include all essential foundation mathematics. As the series title implies, the scripts (of which there are more than 200) described in the text and supplied in code form (available via the internet at http://www.morganclaypool.com/page/isen) will run on both MATLAB and LabVIEW. The text for all volumes contains many examples, and many useful computational scripts, augmented by demonstration scripts and LabVIEW Virtual Instruments (VIs) that can be run to illustrate various signal processing concepts graphically on the user's computer screen. Volume I consists of four chapters that collectively set forth a brief overview of the field of digital signal processing, useful signals and concepts (including convolution, recursion, difference equations, LTI systems, etc), conversion from the continuous to discrete domain and back (i.e., analog-to-digital and digital-to-analog conversion), aliasing, the Nyquist rate, normalized frequency, sample rate conversion, and Mu-law compression, and signal processing principles including correlation, the correlation sequence, the Real DFT, correlation by convolution, matched filtering, simple FIR filters, and simple IIR filters. Chapter 4 of Volume I, in particular, provides an intuitive or "first principle" understanding of how digital filtering and frequency transforms work, preparing the reader for the present volume (Volume II). Volume III of the series covers digital filter design (FIR design using Windowing, Frequency Sampling, and Optimum Equiripple techniques, and Classical IIR design) and Volume IV, the culmination of the series, is an introductory treatment of LMS Adaptive Filtering and applications.

KEYWORDS

Higher-Level Terms:
MATLAB, LabVIEW, MathScript, DSP (Digital Signal Processing), Discrete Time Fourier Transform (DTFT), z-Transform, Discrete Fourier Transform (DFT), Fast Fourier Transform (FFT), Goertzel Algorithm, Discrete Fourier Series (DFS), Frequency Domain, Discrete Frequency Transform
Lower-Level Terms:
FIR, IIR, Direct Form, Direct Form Transposed, Parallel Form, Cascade Form, Lattice Form, Decimation-in-time (DIT)

This volume is dedicated to the memory of the following

Douglas Hunter (1941–1963)
Diane Satterwhite (1949–1961)
John G. Elsberry (1949–1971)
Amelia Megan Au (1967–2007)

Contents

Preface to Volume II

0.1 INTRODUCTION

The present volume is Volume II of the series *DSP for MATLAB*™ *and LabVIEW*™. The entire series consists of four volumes which collectively form a work of twelve chapters that cover basic digital signal processing in a practical and accessible manner, but which nonetheless include essential foundation mathematics. The text is well-illustrated with examples involving practical computation using m-code or MathScript (as m-code is usually referred to in LabVIEW-based literature), and LabVIEW VIs. There is also an ample supply of exercises, which consist of a mixture of paper-and-pencil exercises for simple computations, and script-writing projects having various levels of difficulty, from simple, requiring perhaps ten minutes to accomplish, to challenging, requiring several hours to accomplish. As the series title implies, the scripts given in the text and supplied in code form (available via the internet at http://www.morganclaypool.com/page/isen) are suitable for use with both MATLAB (a product of The Mathworks, Inc.), and LabVIEW (a product of National Instruments, Inc.). Appendix A in each volume of the series describes the naming convention for the software written for the book as well as basics for using the software with MATLAB and LabVIEW.

0.2 THE FOUR VOLUMES OF THE SERIES

The present volume, Volume II of the series, is devoted to discrete frequency transforms. It begins with an overview of a number of well-known continuous domain and discrete domain transforms, and covers the DTFT (Discrete Time Fourier Transform), the DFT (Discrete Fourier Transform), Fast Fourier Transform (FFT), and the z-Transform in detail. Filter realizations (or topologies) are also covered, including Direct, Cascade, Parallel, and Lattice forms.

Volume I of the series, *Fundamentals of Discrete Signal Processing*, consists of four chapters. The first chapter gives a brief overview of the field of digital signal processing. This is followed by a chapter detailing many useful signals and concepts, including convolution, recursion, difference equations, etc. The third chapter covers conversion from the continuous to discrete domain and back (i.e., analog-to-digital and digital-to-analog conversion), aliasing, the Nyquist rate, normalized frequency, conversion from one sample rate to another, waveform generation at various sample rates from stored wave data, and Mu-law compression. The fourth and final chapter of Volume I introduces the reader to many important principles of signal processing, including correlation, the correlation sequence, the Real DFT, correlation by convolution, matched filtering, simple FIR filters, and simple IIR filters.

Volume III of the series, *Digital Filter Design*, covers FIR and IIR design, including general principles of FIR design, the effects of windowing and filter length, characteristics of four types of linear-phase FIR, Comb and MA filters, Windowed Ideal Lowpass filter design, Frequency Sampling design with optimized transition band coefficients, Equiripple FIR design, and Classical IIR design.

Volume IV of the series, *LMS Adaptive Filtering*, begins by explaining cost functions and performance surfaces, followed by the use of gradient search techniques using coefficient perturbation, finally reaching the elegant and computationally efficient Least Mean Square (LMS) coefficient update algorithm. The issues of stability, convergence speed, and narrow-bandwidth signals are covered in a practical manner, with many illustrative scripts. In the second chapter of the volume, use of LMS adaptive filtering in various filtering applications and topologies is explored, including Active Noise Cancellation (ANC),system or plant modeling, periodic component elimination, Adaptive Line Enhancement (ADE), interference cancellation, echo cancellation, and equalization/deconvolution.

0.3   ORIGIN AND EVOLUTION OF THE SERIES

The manuscript from which the present series of four books has been made began with an idea to provide a basic course for intellectual property specialists and engineers that would provide more explanation and illustration of the subject matter than that found in conventional academic books. The idea to provide an accessible basic course in digital signal processing began in the mid-to-late 1990's when I was introduced to MATLAB by Dan Hunter, whose graduate school days occurred after the advent of both MATLAB and LabVIEW (mine did not). About the time I was seriously exploring the use of MATLAB to update my own knowledge of signal processing, Dr. Jeffrey Gluck began giving an in-house course at the agency on the topics of convolutional coding, trellis coding, etc., thus inspiring me to do likewise in the basics of DSP, a topic more in-tune to the needs of the unit I was supervising at the time. Two short courses were taught at the agency in 1999 and 2000 by myself and several others, including Dr. Hal Zintel, David Knepper, and Dr.Pinchus Laufer. In these courses we stressed audio and speech topics in addition to basic signal processing concepts. Thanks to The Mathworks, Inc., we were able to teach the in-house course with MATLAB on individual computers, and thanks to Jim Dwyer at the agency, we were able to acquire several server-based concurrent-usage MATLAB licenses, permitting anyone at the agency to have access to MATLAB. Some time after this, I decided to develop a complete course in book form, the previous courses having consisted of an ad hoc pastiche of topics presented in summary form on slides, augmented with visual presentations generated by custom-written scripts for MATLAB. An early draft of the book was kindly reviewed by Motorola Patent Attorney Sylvia Y. Chen, which encouraged me to contact Tom Robbins at Prentice-Hall concerning possible publication. By 2005, Tom was involved in starting a publishing operation at National Instruments, Inc., and introduced me to LabVIEW with the idea of possibly crafting a book on DSP to be compatible with LabVIEW. After review of a manuscript draft by a panel of three in early 2006, it was suggested that all essential foundation mathematics be included so the book would have both academic and professional appeal. Fortunately, I had long since retired from the agency and was able to devote the considerable amount of time needed for such a project. The result is a book suitable for use in both academic and professional settings, as it includes essential mathematical formulas and concepts as well as simple or "first principle" explanations that help give the reader a gentler entry into the more conventional mathematical treatment.

This double-pronged approach to the subject matter has, of course, resulted in a book of considerable length. Accordingly, it has been broken into four modules or volumes (described above) that together form a comprehensive course, but which may be used individually by readers who are not in need of a complete course.

Many thanks go not only to all those mentioned above, but to Joel Claypool of Morgan & Claypool, Dr. C.L. Tondo and his troops, and, no doubt, many others behind the scenes whose names I have never heard, for making possible the publication of this series of books.

Forester W. Isen
December 2008

CHAPTER 1

The Discrete Time Fourier Transform

1.1 OVERVIEW

1.1.1 IN THE PREVIOUS VOLUME

The previous volume of the series, Volume I, covered DSP fundamentals such as basic signals and LTI systems, difference equations, sampling, the Nyquist rate, normalized frequency, correlation, convolution, the real DFT, matched filtering, and basic IIR and FIR filters.

1.1.2 IN THIS VOLUME

In this volume, Volume II of the series, we take up discrete frequency transforms in detail, including an overview of many transforms, both continuous-domain and discrete-domain, followed in sequence by detailed discussions of a number of discrete transforms, knowledge of which is generally deemed essential in the signal processing field.

1.1.3 IN THIS CHAPTER

We are now prepared in this chapter to begin a detailed exploration of discrete frequency transforms. A number of such transforms exist, and we'll begin by summarizing all of the basic facts and comparing each to the others to better emphasize the important characteristics of each distinct transform. We include a brief mention of continuous signal domain transforms for background and perspective, but concentrate most of our effort on discrete signal transforms. All of the transforms we'll investigate, both the continuous and the discrete domain types, work on the same general concept–summing (or integrating in the case of continuous time signals) the product of the signal and orthogonal-pair correlating waveforms of different frequencies.

The transforms covered in detail in this book section are the **Discrete Time Fourier Transform (DTFT)**, which is covered in this chapter, the z-**Transform (z-T)**, covered in the following chapter, and the **Discrete Fourier Transform (DFT)**, covered in the third and final chapter of this volume.

By the end of this chapter, the reader will have learned how to evaluate the frequency response of an LTI system using the DTFT. This sets the stage for the following chapter, which discusses the more generalized z-transform, which is in widespread use in industry and academia as the standard method to describe the transfer function of an LTI system. This will be followed in the succeeding chapter by a detailed look at the workhorse of practical frequency domain work, the DFT and a fast implementation, the decimation-in-time FFT, as well as time domain convolution using the frequency domain, the Goertzel algorithm, computing the DTFT using the DFT, etc.

1.2   SOFTWARE FOR USE WITH THIS BOOK

The software files needed for use with this book (consisting of m-code (.m) files, VI files (.vi), and related support files) are available for download from the following website:

http://www.morganclaypool.com/page/isen

The entire software package should be stored in a single folder on the user's computer, and the full file name of the folder must be placed on the MATLAB or LabVIEW search path in accordance with the instructions provided by the respective software vendor (in case you have encountered this notice before, which is repeated for convenience in each chapter of the book, the software download only needs to be done once, as files for the entire series of four volumes are all contained in the one downloadable folder).

See Appendix A for more information.

1.3   INTRODUCTION TO TRANSFORM FAMILIES

The chief differences among the transforms mentioned below involve whether they 1) operate on continuous or discrete time signals, 2) provide continuous or discrete frequency output, and 3) use constant unity-amplitude correlators (in the case of the Fourier family of transforms), or dynamic (decaying, steady-state, or growing correlators) in the case of the Laplace and z- transforms.

The following table summarizes the main characteristics of a number of well-known transforms with respect to the following categories: Input Signal Domain (continuous **C** or discrete **D** signal), Output (Frequency) Domain (produces continuous **C** or discrete **D** frequency output), and Correlator Magnitude (constant, unity magnitude for Fourier-based transforms, or variable magnitudes (decaying, growing, or constant unity) for Laplace Transform and z-Transform, accordingly as $e^{-\sigma t}$ or $|z|^n$, respectively.

Transform	Input	Output	Correlator Mag.
Laplace Transform	C	C	$e^{-\sigma t}$
Fourier Transform	C	C	1
Fourier Series	C	D	1
Discrete Time Fourier Transform	D	C	1
Discrete Fourier Series	D	D	1
Discrete Fourier Transform	D	D	1
z-transform	D	C	$|z|^n$

For purposes of discrete signal processing, what is needed is a numerically computable representation (transform) of the input sequence; that is to say, a representation which is itself a finite but complete representation that can be used to reenter the time domain, i.e., reconstruct the original signal. For transforms that are not computable in this sense, samples of the transform can be computed. Of all the transforms discussed in the following section of the chapter, only the **Discrete Fourier Series (DFS)** and Discrete Fourier Transform (DFT) are computable transforms in the sense mentioned above.

The use of dynamic correlators results in a transform that is an algebraic expression that implicitly or explicitly contains information on the system poles and zeros. The system response to signals other than steady-state, unity amplitude signals can readily be determined, although such transforms can also be evaluated to produce the same result provided by the Fourier Transform (in the case of the Laplace Transform) or the DTFT, DFS, and DFT (in the case of the z-Transform). Thus, the Laplace Transform and z-Transform are more generalized transforms having great utility for system representation and computation of response to many types of signals from the continuous and discrete time domains, respectively.

While this book is concerned chiefly with discrete signal processing, we give here a brief discussion of certain continuous time transforms (Laplace, Fourier, Fourier Series) to serve as background or points of reference for the discrete transforms that will be discussed in more detail below and in chapters to follow.

1.3.1 FOURIER FAMILY (CONSTANT UNITY-MAGNITUDE CORRELATORS)
Fourier Transform

$$F(\omega) = \int_{-\infty}^{\infty} x(t)e^{-j\omega t} dt$$

The Fourier transform operates on continuous time, aperiodic signals and evaluates the frequency response in the continuous frequency domain. The correlators are complex exponentials having constant unity amplitude. Both t (time) and ω (frequency) run from negative infinity to positive infinity. The Fourier Transform is a reversible transform; the inverse transform is

$$x(t) = \int_{-\infty}^{\infty} F(\omega)e^{j\omega t} d\omega$$

Fourier Series
Many signals of interest are periodic, that is, they are composed of a harmonic series of cosines and sines. For a periodic, continuous time signal of infinite extent in time, a set of coefficients can be obtained based on a single period (between times t_o and $t_o + T$) of the signal $x(t)$:

$$c_k = \frac{1}{T} \int_{t_o}^{t_o+T} x(t)e^{-jk\omega_0 t} dt$$

where T is the reciprocal of the fundamental frequency F_0 and $k = 0, \pm1, \pm2, ...$

For real $x(t)$, c_k and c_{-k} are complex conjugates. If we say

$$c_k = |c_k|\, e^{j\theta_k}$$

then the original sequence can be reconstructed according to the formula

$$x(t) = c_0 + 2 \sum |c_k| \cos(2\pi k F_0 t + \theta_k)$$

An equivalent expression is

$$x(t) = a_0 + \sum_{k=1}^{\infty} a_k \cos(2\pi k F_0 t) - b_k \sin(2\pi k F_0 t)$$

where $a_0 = c_0$, $a_k = 2\,|c_k| \cos\theta_k$, and $b_k = 2\,|c_k| \sin\theta_k$.

Discrete Time Fourier Transform (DTFT)
The DTFT is defined for the discrete time input signal $x[n]$ as

$$DTFT(x[n]) = X(e^{j\omega}) = \sum_{n=-\infty}^{\infty} x[n]e^{-j\omega n} \tag{1.1}$$

where ω (radian frequency) is a continuous function and runs from $-\pi$ to π, and $x[n]$ is absolutely summable, i.e.,

$$\sum_{n=-\infty}^{\infty} |x[n]| < \infty$$

The inverse DTFT (IDTFT) is defined as

$$x[n] = \frac{1}{2\pi} \int_{-\pi}^{\pi} X(e^{j\omega})e^{j\omega n} d\omega \tag{1.2}$$

DTFT theory will be discussed in detail below, while computation of the DTFT using the DFT will be discussed in the chapter on the DFT.

Discrete Fourier Series (DFS)
A periodic sequence $x[n]$ ($-\infty < n < \infty$) may be decomposed into component sequences that comprise a harmonic series of complex exponentials. Since a sampled sequence is bandlimited, it follows that the frequency of the highest harmonic is limited to the Nyquist rate. The normalized frequencies of the harmonics are $2\pi k/N$, with $k = 0{:}1{:}N\text{-}1$. Since the transform involves a finite number of frequencies to be evaluated, the DFS is a computable transform.
The DFS coefficients $\widetilde{X}[k]$ are

$$DFS(x[n]) = \widetilde{X}[k] = \sum_{n=0}^{N-1} \widetilde{x}[n]e^{-j2\pi kn/N} \tag{1.3}$$

where k is an integer and $\tilde{x}[n]$ is one period of the periodic sequence $x[n]$. Typical ranges for k are: 0:1:N-1, or -N/2+1:1:N/2 for even length sequences, or -(N-1)/2:1:(N-1)/2 for odd length sequences.

Once having the DFS coefficients, the original sequence $x[n]$ can be reconstructed using the following formula:

$$x[n] = \frac{1}{N} \sum_{k=0}^{N-1} \tilde{X}[k] e^{j2\pi kn/N} \tag{1.4}$$

The DFS, a computable transform, forms an important theoretical basis for the Discrete Fourier Transform and will be discussed in more detail in the chapter on the DFT.

Discrete Fourier Transform (DFT)

$$DFT[k] = \frac{1}{N} \sum_{n=0}^{N-1} x[n] e^{-j2\pi kn/N}$$

The DFT (a computable transform) operates on discrete, or sampled signals, and evaluates frequency response at a number (equal to about half the sequence length) of unique frequencies. The correlators are complex exponentials having constant, unity amplitude, and the frequencies k range from 0 to $N-1$ or, for even-length sequences, $-N/2+1$ to $N/2$ and $-N/2$ to $-N/2$ for odd-length DFTs.

The DFT (including its efficient implementation, the FFT) will be discussed extensively in the chapter that follows the chapter on the z-transform.

1.3.2 LAPLACE FAMILY (TIME-VARYING-MAGNITUDE CORRELATORS)
Laplace Transform (LT)
The LT is defined as

$$\pounds(s) = \int_{-\infty}^{\infty} x(t)e^{-st} dt = \int_{-\infty}^{\infty} x(t)e^{-\sigma t} e^{-j\omega t} dt$$

The LT is the standard frequency transform for use with continuous time domain signals and systems. The parameter s represents the complex number $\sigma + j\omega$, with σ, a real number, being a damping coefficient, and $j\omega$, an imaginary number, representing frequency. Both σ and ω run from negative infinity to positive infinity. The correlators generated by e^{-st} are complex exponentials having amplitudes that decay, grow, or retain unity-amplitude over time, depending on the value of σ. By varying both σ and ω, the poles and zeros of the signal or system can be identified. Results are graphed in the s-Domain (the complex plane), using rectangular coordinates with σ along the horizontal axis, and $j\omega$ along the vertical axis. The magnitude of the transform can be plotted along a third dimension in a 3-D plot if desired, but, more commonly, a 2-D plot is employed showing only the locations of poles and zeros.

The LT is a reversible transform, and can be used to solve differential equations, such as those representing circuits having inductance and capacitance, in the frequency domain. The time domain solution is then obtained by using the Inverse LT. The LT is used extensively for circuit analysis and representation in the continuous domain. We'll see later in the book that certain well known or classical IIRs (Butterworth, Chebyshev, etc) have been extensively developed in the continuous domain using LTs, and that the Laplace filter parameters can be converted to the digital domain to create an equivalent digital IIR.

Note that the FT results when $\sigma = 0$ in the Laplace transform. That is to say, when the damping coefficient is zero, the Laplace correlators are constant, unity-amplitude complex exponentials just as those of the FT. Information plotted along the Imaginary axis in the s- or Laplace domain is equivalent, then, to the FT.

z-Transform (**z-T**)

The z-transform (z-T) is a discrete time form of the Laplace Transform. For those readers not familiar with the LT, study of the z-T can prove helpful since many Laplace properties and transforms of common signals are analogous to those associated with the z-T. The z-T is defined as

$$X(z) = \sum_{n=-\infty}^{\infty} x[n]z^{-n}$$

The z-T converts a number sequence into an algebraic expression in z, and, in the reverse or inverse z-T, converts an algebraic expression in z into a sequence of numbers. The z-T is essentially a discrete-time version of the Laplace transform. The correlators are discrete-time complex exponentials with amplitudes that grow, shrink, or stay the same according to the value of z (a complex number) at which the transform is evaluated. Values of z having magnitudes < 1 result in a correlation of the signal with a decaying (discrete) complex exponential, evaluation with z having a magnitude equal to 1 results in a Fourier-like response, and evaluation with z having a magnitude greater than 1 results in a correlation of the signal with a growing discrete complex exponential. Results can be plotted in the z-Domain, which is the complex plane using polar coordinates of r and θ, where θ corresponds to normalized radian frequency and r to a damping factor. The unit circle in the z-domain corresponds to the imaginary axis in the s-Domain; the left hand s-plane marks a region of stable pole values in the s-plane that corresponds to the area inside the unit circle in the z-plane. As in the s-plane, the z-T magnitude may be plotted using a 3-D plot, or, using a 2-D plot, only the location of the poles and zeros may be plotted.

Among the discrete-signal transforms, the z-T is more general than the DTFT, since in the z-T, the test exponentials may also have decay factors (negative or positive). All of the members of the Fourier transform family use exponentials of constant unity amplitude to perform their correlations with the signal of unknown frequency content, and thus cannot give pole and zero locations as can the Laplace and z- transforms.

The z-transform will be discussed extensively in the next chapter.

Additional Transforms

There are also several other transforms that are derived from the DFT; namely, the Discrete Cosine Transform (DCT), and the Discrete Sine Transform (DST). These transforms use multiples of half-cycles of either the cosine or sine, respectively, as the correlators, rather than multiples of full cycles as in the DFT. The bin values are real only. A form of the DCT called the MDCT (Modified DCT) is used in certain audio compression algorithms such as MP3.

Reference [1] gives a thorough and very accessible development of the DFT, and also briefly discusses the Laplace transform, the z-transform, the DCT, and the DST.

1.4   THE DTFT

The DTFT provides a continuous frequency spectrum for a sampled signal, as opposed to a discrete frequency spectrum (in which only a finite number of frequency correlators are used, as is true of the DFT and DFS).

In Eq. (1.1), $x[n]$ is a sampled signal which may be of either finite or infinite extent, and ω, a continuous function of frequency which may assume values from $-\pi$ to π.

Example 1.1.   Derive an algebraic expression for the DTFT of the function $0.9^n u[n]$, and then evaluate it numerically at frequencies between 0 and π radians, at intervals of 0.01 radian.

We must evaluate the expression

$$DTFT(x[n]) = \sum_{n=-\infty}^{\infty} x[n]e^{-j\omega n}$$

We note that $x[n] = 0$ for $n < 0$, that $x[n]$ itself is absolutely summable since it is a decreasing geometric series. The n-th term of the summation is

$$0.9^n e^{-j\omega n}$$

and we note that each successive term is arrived at by multiplying the previous term by the number $0.9e^{-j\omega}$ and thus the net sequence forms a geometric series the sum of which is

$$\frac{1}{(1 - 0.9e^{-j\omega})} \tag{1.5}$$

We can evaluate this expression at a finite number of values of ω and plot the result. The following code directly evaluates expression (1.5) at frequencies between 0 and π radians, as specified by the vector w:

```
w = 0:0.01:pi; DTFT = 1./(1-0.9*exp(-j*(w)));
figure(8); plot(w/(pi),abs(DTFT));
xlabel('Normalized Frequency'); ylabel('Magnitude')
```

Example 1.2.   Evaluate and plot the magnitude of the DTFT of the following sequence: $[1, 0, 1]$.

$$F(\omega) = \sum_{n=0}^{2} x[n]e^{-j\omega n} = \sum [1, 0, 1][e^{-j\omega 0}, e^{-j\omega 1}, e^{-j\omega 2}] = 1 + e^{-j\omega 2} \qquad (1.6)$$

From our earlier work, we recognize the impulse response $[1,0,1]$ as that of a simple bandstop filter. We can show that this is so by evaluating Eq. (1.6) at a large (but necessarily finite) number of values of ω with the following code, the results of which are shown in Fig. 1.1.

```
w = 0:0.01:pi; DTFT = 1+exp(-j*2*w);
figure(8); plot(w/(pi),abs(DTFT));
xlabel('Normalized Frequency'); ylabel('Magnitude')
```

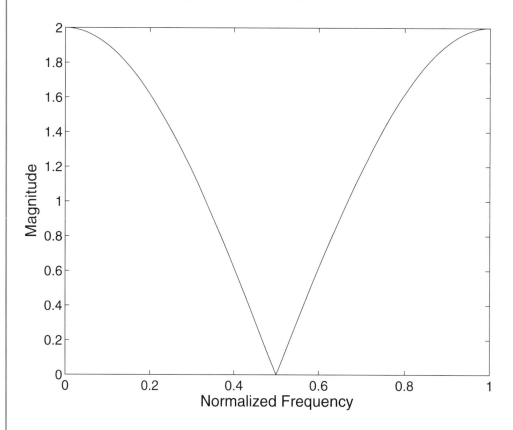

Figure 1.1: Magnitude of the DTFT of the simple notch filter $[1, 0, 1]$.

Example 1.3.   Write a script that will evaluate and plot the magnitude and phase of the DTFT for any sequence; test it for the sequences $[1, 0, 1]$, $[1, 0, -1]$, and $[1, 0, 0, 1]$.

Such a script should allow one to specify how many samples M of the DTFT to compute over the interval 0 to $R\pi$, with $R = 1$ being suitable for real $x[n]$ and $R = 2$ being suitable for complex $x[n]$. Values of R greater than 2 allow demonstration of periodicity of the DTFT. The code creates an n-by-k matrix $dMat$ of complex correlators, where each column is a complex correlator of length n and frequency k. The DTFT is obtained by multiplying the signal vector x on the right by $dMat$. Each element in the resulting row vector of frequency responses is obtained as the inner or dot product of the signal vector x with a column of $dMat$.

```
function LV_DTFT_Basic(x,M,R)
% LV_DTFT_Basic([1,0,1],300,1)
N = length(x); W = exp(-j*R*pi/M); k = 0:1:M-1;
n = 0:1:N-1; dMat = W.^(n'*k); d = x*dMat; figure(9)
subplot(2,2,1); plot(R*[0:1:M-1]/M,abs(d));
grid; xlabel('Norm Freq'); ylabel('Mag')
subplot(2,2,2); plot(R*[0:1:M-1]/M,angle(d))
grid; xlabel('Norm Freq'); ylabel('Radians')
subplot(2,2,3); plot(R*[0:1:M-1]/M,real(d));
grid; xlabel('Norm Freq'); ylabel('Real')
subplot(2,2,4); plot(R*[0:1:M-1]/M,imag(d))
grid; xlabel('Norm Freq'); ylabel('Imag')
```

The result from making the call

$$\text{LV\_DTFT\_Basic}([1,0,1],300,1)$$

is shown in Fig. 1.2.

A more versatile version of the above code is the script (see exercises below)

$$LVxDTFT(x, n, M, R, FreqOpt, FigNo)$$

which, from sequence x having time indices n, computes M frequency samples over the interval $R\pi$, which can be computed symmetrically or asymmetrically with respect to frequency zero ($FreqOpt$ = 1 for symmetrical, 2 for asymmetrical). The radian frequencies of evaluation would be, for the asymmetrical option

$$R\pi([0 : 1 : M - 1])/M$$

and for the symmetrical option

$$R\pi([-(M - 1)/2 : 1 : (M - 1)/2])/M \quad (M \text{ odd})$$

$$R\pi([-M/2 + 1 : 1 : M/2])/M \quad (M \text{ even})$$

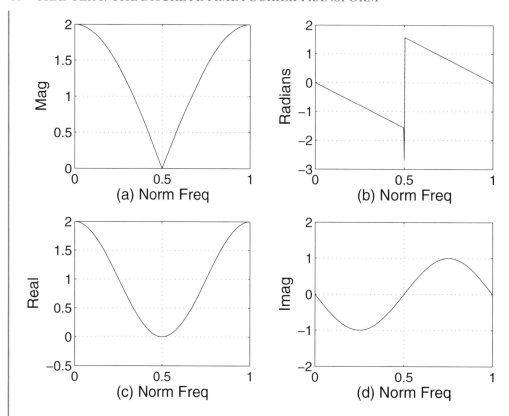

Figure 1.2: (a) Magnitude of DTFT of the sequence [1 0 1]; (b) Phase response of DTFT; (c) Real component of DTFT ; (d) Imaginary component of DTFT.

The desired figure number is supplied as *FigNo*, an option allowing you to create different figures for comparison with each other. A typical call is

LVxDTFT([1,0,1],[0:1:2],300,2,1,88)

which results in Fig. 1.3.

A second script (for a complete description of input arguments, see exercises below)

$$LVxDTFT\_MS(x, SampOffset, FreqOffsetExp, M, R, TimeOpt, FreqOpt)$$

allows you to enter one sequence, and the second sequence is created as a modification of the first, delayed by *SampOffset* samples and offset in frequency by the complex exponential *FreqOffsetExp*. Input arguments *M* and *R* are as described for the script *LVxDTFT*; *FreqOpt* determines whether the DTFT is computed symmetrically or asymmetrically about frequency zero, as described above

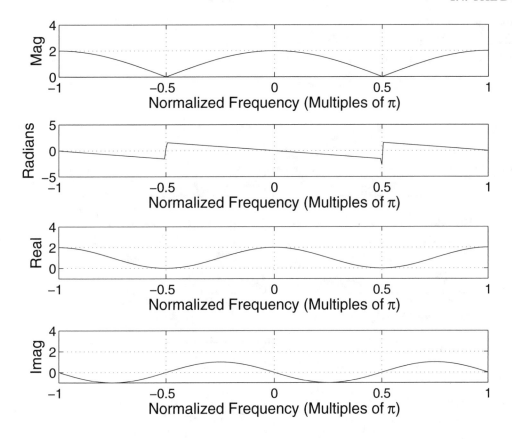

Figure 1.3: (a) Magnitude of DTFT of the sequence [1,0,1]; (b) Phase of DTFT of same; (c) Real part of DTFT of same; (d) Imaginary part of DTFT of same.

for the script *LVxDTFT*. For the asymmetrical time option (determined by the input argument *TimeOpt*), the sequence time indices of the first sequence are given as

$$n = 0 : 1 : N - 1;$$

where N is the length of x. For the symmetrical time index option, the time indices are given as

$$n = -(N-1)/2 : 1 : (N-1)/2 \quad (n \text{ odd})$$

$$n = -N/2 + 1 : 1 : N/2 \quad (n \text{ even})$$

This script is useful for demonstrating the effect on the DTFT of time and frequency shifts to a first test sequence. A typical call, which results in Fig. 1.4, is

```
nN = (0:1:100)/100;
LVxDTFT_MS([cos(2*pi*25*nN)],0,...
exp(j*2*pi*12.5*nN),200,2,2,1)
```

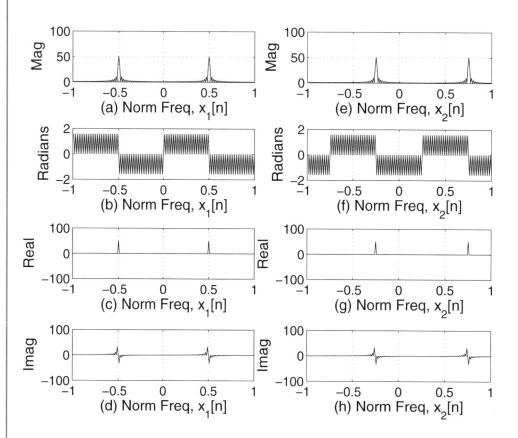

Figure 1.4: DTFT of first sequence, with its magnitude and phase and real and imaginary parts being shown, respectively, in plots (a)-(d); DTFT of second sequence, which is the first sequence offset in frequency by $\pi/4$ radian, with its magnitude, phase, and real and imaginary parts being shown, respectively, in plots (e)-(h). All frequencies are normalized, i.e., in units of π radians.

We will make use of these scripts shortly while studying the various properties of the DTFT.

1.5   INVERSE DTFT

The Inverse DTFT, i.e., the original time domain sequence $x[n]$ from which a given DTFT was produced, can be reconstructed by evaluating the following integral:

$$x[n] = \frac{1}{2\pi} \int_{-\pi}^{\pi} X(e^{j\omega}) e^{j\omega n} d\omega \tag{1.7}$$

We will illustrate this with several examples, one analytic and the other numerical.

Example 1.4.   Using Eq. (1.7), compute $x[0]$, $x[1]$, and $x[2]$ from the DTFT obtained in Eq. (1.6).

We get

$$x[0] = \frac{1}{2\pi} \int_{-\pi}^{\pi} (1 + e^{-j\omega 2}) e^{j\omega 0} d\omega$$

and

$$x[1] = \frac{1}{2\pi} \int_{-\pi}^{\pi} (1 + e^{-j\omega 2}) e^{j\omega 1} d\omega$$

and

$$x[2] = \frac{1}{2\pi} \int_{-\pi}^{\pi} (1 + e^{-j\omega 2}) e^{j\omega 2} d\omega$$

The formula for $x[0]$ reduces to

$$\frac{1}{2\pi} \int_{-\pi}^{\pi} (1 + e^{-j\omega 2}) d\omega = \frac{1}{2\pi} (\int_{-\pi}^{\pi} d\omega + \int_{-\pi}^{\pi} e^{-j\omega 2} d\omega)$$

which is

$$\frac{1}{2\pi} (\omega \mid_{-\pi}^{\pi} + \int_{-\pi}^{\pi} e^{-j\omega 2} d\omega) = 1 + 0 = 1$$

where we note that

$$\int_{-\pi}^{\pi} e^{\pm j\omega n} d\omega = \begin{cases} 2\pi & \text{if} \quad n = 0 \\ 0 & \text{if} \quad n = \pm 1, \pm 2... \end{cases}$$

For $x[1]$ we get

$$x[1] = \frac{1}{2\pi} \int_{-\pi}^{\pi} (1 + e^{-j\omega 2}) e^{j\omega 1} d\omega = \frac{1}{2\pi} \int_{-\pi}^{\pi} (e^{j\omega 1} + e^{-j\omega 1}) d\omega = 0$$

The formula for $x[2]$ is the same as that for $x[0]$ with the exception of the sign of the complex exponential, which does not affect the outcome. The reconstructed sequence is therefore $[1,0,1]$, as expected.

Example 1.5.   Using numerical integration, approximate the IDTFT that was determined analytically in the previous example.

Let's reformulate the code to obtain the DTFT from -pi to +pi, and to use a much finer sample spacing (this will improve the approximation to the true, continuous spectrum DTFT), and then perform the IDTFT one sample at a time in accordance with Eq. (1.2):

```
N=10^3; dw = 2*pi/N; w = -pi:dw:pi*(1-2/N);
DTFT = 1+exp(-j*2*w);
x0 = (1/(2*pi))*sum(DTFT.*exp(j*w*0)*dw)
x1 = (1/(2*pi))*sum(DTFT.*exp(j*w*1)*dw)
x2 = (1/(2*pi))*sum(DTFT.*exp(j*w*2)*dw)
```

Running the preceding code yields the following answer, which rounds to the original sequence, $[1,0,1]$:

```
x0 = 1.0000 - 0.0000i
x1 = -4.9127e-017
x2 = 1.0000 + 0.0000i
```

1.6   A FEW PROPERTIES OF THE DTFT

1.6.1   LINEARITY

The DTFT of a linear combination of two sequences $x_1[n]$ and $x_2[n]$ is equal to the sum of the individual responses, i.e.,

$$DTFT(ax_1[n] + bx_2[n]) = aDTFT(x_1[n]) + bDTFT(x_2[n])$$

1.6.2   CONJUGATE SYMMETRY FOR REAL $x[n]$

For real $x[n]$, the real part of the DTFT shows even symmetry ($X(e^{j\omega}) = X(e^{-j\omega})$), and the imaginary part shows odd symmetry ($X(e^{j\omega}) = -X(e^{-j\omega})$).

1.6.3   PERIODICITY

The DTFT of a sequence $x[n]$ repeats itself every 2π:

$$X(e^{jw}) = X(e^{j(w+2\pi n)})$$

where $n = 0, \pm 1, \pm 2...$

To illustrate this principle, consider the following: for a given sequence length, the Nyquist limit is half the sequence length, and this represents a frequency shift of π radians. To shift 2π

radians therefore is to shift by a frequency equal to the sequence length. The following m-code, the results of which are illustrated in Fig. 1.5, verifies this property.

SR = 100; nN = (0:1:SR)/SR;
LVxDTFT_MS([cos(2*pi*25*nN)],0,exp(j*2*pi*SR*nN),200,2,2,1)

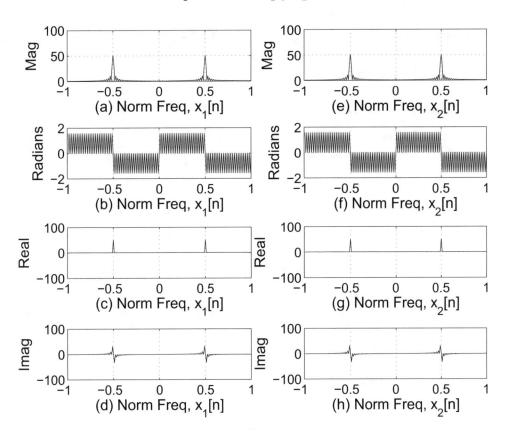

Figure 1.5: DTFT of first sequence, with its magnitude and phase and real and imaginary parts being shown, respectively, in plots (a)-(d); DTFT of second sequence, which is the first sequence offset in frequency by 2π radians, with its magnitude, phase, and real and imaginary parts being shown, respectively, in plots (e)-(h). All frequencies are normalized, i.e., in units of π radians.

You can gain insight by noting that the code

$$nN = (0:1:100)/100; y = exp(j*2*pi*100*nN)$$

yields $y = ones(1, 101)$, which clearly transforms the original sequence into itself, i.e., the new sequence is the same as the old, and hence the DTFT is the same. In other words, the DTFT of a sequence repeats itself for every frequency shift of 2π radians of the original sequence.

1.6.4   SHIFT OF FREQUENCY

If the signal $x[n]$ is multiplied by a complex exponential of frequency F_0, the result is that the DTFT of $x[n]$ is shifted.

$$DTFT(x[n]e^{j\omega_0 n}) = X(e^{j(\omega-\omega_0)})$$

To demonstrate this property, we can use the script *LVxDTFT_MS*. We pick the short sequence [1,0,1] as $x[n]$, and specify no sample offset, but a frequency offset of $2\pi/16$ radians. We thus make the call

LVxDTFT_MS([1,0,1],0,exp(j*2*pi*1*(0:1:2)/16),100,2,1,2)

which results in Fig. 1.6. The reader should be able to verify by visual comparison of plots (a) and (e) that the magnitude of frequency response has in fact been shifted by $\pi/8$ radian.

1.6.5   CONVOLUTION

The DTFT of the time domain convolution of two sequences is equal to the product of the DTFTs of the two sequences.

$$DTFT(x_1[n] * x_1[n]) = X_1(e^{j\omega})X_2(e^{j\omega})$$

Example 1.6.   Consider the two sequences [1, 0, 1] and [1, 0, 0, −1]. Obtain the time domain convolution by taking the inverse DTFT of the product of the DTFTs of each sequence, and confirm the result using time domain convolution.

The DTFT of the first sequence is

$$X_1(e^{j\omega}) = 1 + e^{-j\omega 2}$$

and for the second sequence

$$X_2(e^{j\omega}) = 1 - e^{-j\omega 3}$$

The product is

$$1 + e^{-j\omega 2} - e^{-j\omega 3} - e^{-j\omega 5}$$

or stated more completely

$$1e^{-j\omega 0} + 0e^{-j\omega 1} + e^{-j\omega 2} - e^{-j\omega 3} + 0e^{-j\omega 4} - e^{-j\omega 5}$$

which can be recognized by inspection as the DTFT of the time domain sequence

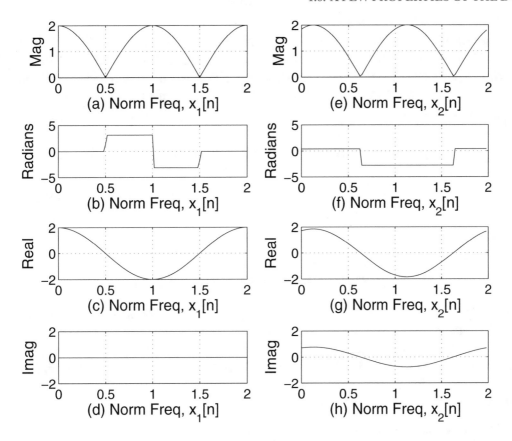

Figure 1.6: DTFT of first sequence, with its magnitude and phase and real and imaginary parts being shown, respectively, in plots (a)-(d); DTFT of second sequence, which is the first sequence offset in frequency by $\pi/8$ radian, with its magnitude, phase, and real and imaginary parts being shown, respectively, in plots (e)-(h). All frequencies are normalized, i.e., in units of π radians.

$$[1, 0, 1, -1, 0, -1]$$

To confirm, we make the following call:

$$\mathbf{y = conv([1,0,1],[1,0,0,-1])}$$

which produces the same result.

1.6.6   EVEN AND ODD COMPONENTS

If a real sequence $x[n]$ is decomposed into its even and odd components, and the DTFT taken of each, it will be found that the DTFT of the even part is all real and equal to the real part of $DTFT(x[n])$, while the DTFT of the odd part will be found to be imaginary only and equal to the imaginary part of $DTFT(x[n])$. Using $X(e^{j\omega})$, $XE(e^{j\omega})$, $XO(e^{j\omega})$ as the DTFT's of $x[n]$, $xe[n]$, and $xo[n]$, we would have

$$\text{Re}(X(e^{j\omega})) = \text{Re}(XE(e^{j\omega})) = XE(e^{j\omega})$$

and

$$\text{Im}(X(e^{j\omega})) = \text{Re}(XO(e^{j\omega})) = XO(e^{j\omega})$$

Example 1.7.   Write a script that will demonstrate the above even-odd properties.

We can perform the even-odd decomposition using the script *LVEvenOddSymmZero* presented earlier, and then use the script *LVxDTFT* in three separate calls to open three separate windows to show $X(e^{j\omega})$, $XE(e^{j\omega})$, and $XO(e^{j\omega})$; the following script also reconstructs the original DTFT as the sum of its even and odd components, subtracts this from the DTFT of the original signal, and obtains the RMS error, which should prove to be essentially zero, within the limits of roundoff error.

```
x = [1:1:9]; [xe,xo,m] = LVEvenOddSymmZero(x,[0:1:8]);
d = LVxDTFT(x,[0:1:8],200,2,2,10);
de = LVxDTFT(xe,m,200,2,2,11);
do = LVxDTFT(xo,m,200,2,2,12);
RMS = sqrt((1/200)*sum((d - (de+do)).^2))
```

The value of RMS reported after running the above code was [1.0190e-016, -1.6063e-015i], which is essentially zero, within the limits of roundoff error.

1.6.7   MULTIPLICATION BY A RAMP

$$DTFT(nx[n]) = j\frac{dX(e^{j\omega})}{d\omega}$$

This property states that if the DTFT of $x[n]$ is $X(e^{j\omega})$, then the DTFT of $nx[n]$ is the derivative with respect to ω of $X(e^{j\omega})$ multiplied by j.

1.7   FREQUENCY RESPONSE OF AN LTI SYSTEM

1.7.1   FROM IMPULSE RESPONSE

An interesting and useful result occurs by convolving a complex exponential $e^{j\omega k}$ of radian frequency of $\omega = \omega_0$ with an LTI system having an impulse response represented by $h[n]$:

$$y[k] = \sum_{n=-\infty}^{\infty} h[n]e^{j\omega_0(k-n)} = \left(\sum h[n]e^{-j\omega_0 n}\right)e^{j\omega_0 k} \tag{1.8}$$

The rightmost expression in Eq. (1.8) is the input signal $e^{j\omega_0 k}$ scaled by the DTFT of $h[n]$ evaluated at ω_0. Since the DTFT evaluated at a single frequency is a complex number, it is sometimes convenient to represent it as a magnitude and phase angle. Thus, for each frequency ω_0 we would have the result

$$y[k] = H(e^{j\omega_0})e^{j\omega_0 k} = (|H(e^{j\omega_0})|)(\angle H(e^{j\omega_0}))e^{j\omega_0 k} \tag{1.9}$$

An interpretation of this result is that a sinusoidal excitation of frequency ω_0 to an LTI system produces an output that is a sinusoid of the same frequency, scaled by the magnitude of the DTFT of $h[n]$ at ω_0, and phase shifted by the angle of the DTFT of $h[n]$ evaluated at ω_0. This is consistent with the principle of Sinusoidal Fidelity discussed in Volume I of this series.

Example 1.8.   Consider the LTI system whose impulse response is [1, 0, 1]. Determine the magnitude of the response to the complex exponential.

$$\exp(j*2*pi*(0:1:31)*5/32)$$

We note that the normalized frequency is $5/16 = 0.3125$ and obtain the magnitude of H at the given frequency as 1.111 using the following code:

w = 5*pi/16; DTFT = 1+exp(-j*2*w);
magH = abs(DTFT)

To obtain the magnitude of response via time domain convolution, we make the following call:

tdMagResp = max(abs(conv([1,0,1],exp(j*2*pi*(0:1:31)*5/32))))

which produces the identical result, 1.111.

Equation (1.9) may be generalized and applied to real sinusoids. Thus, the **steady state** response $y[n]$ of an LTI system to a cosine (or sine) of magnitude A, frequency ω_0 and phase angle ϕ, with the DTFT magnitude and angle being M and θ is

$$y[n] = MA\cos(\omega_0 n + \phi + \theta) \tag{1.10}$$

Example 1.9.   Verify Eq. (1.10) using both cosine and sine waves having normalized frequencies of 0.53 for the LTI system whose impulse response is [1, 0, 1].

We'll use a discrete cosine of the stated frequency with $\phi = 0$ and $M = 1$, and then compute the output according to Eq. (1.10) and then by convolution with the impulse response. The plot will show that the results are identical except for the first two and last two samples (recall that Eq. (1.10) represents a steady state response). In the code below, you can substitute the sine function (*sin*) for the cosine function (*cos*) as well as change the frequency F and test signal length N.

> **F = 0.53; N = 128; w = F*pi; dtft = 1+exp(-j*2*w);**
> **M = abs(dtft); theang = angle(dtft);**
> **y1 = M*cos(2*pi*(0:1:N-1)*(N/2*F)/N + theang);**
> **y2 = conv([1 0 1],cos(2*pi*(0:1:N-1)*(N/2*F)/N));**
> **stem(y1,'bo'); hold on; stem(y2,'r*')**

Since LTI systems obey the law of superposition, Eq. (1.10) can be generalized for an input signal comprising a sum or superposition of sinusoids:

$$y[n] = \sum_{k=1}^{K} M_k A_k \cos(\omega_k n + \phi_k + \theta_k) \tag{1.11}$$

1.7.2 FROM DIFFERENCE EQUATION

Thus, far we have been using the impulse response of an LTI system to determine the system's frequency response; in many cases, a difference equation may be what is immediately available. While it is certainly possible to process a unit impulse using the difference equation and thus obtain the impulse response, it is possible to obtain the frequency response directly from the difference equation. Assuming that the difference equation is of the form

$$y[n] + \sum_{k=1}^{K} a_k y[n-k] = \sum_{m=0}^{M} b_m x[n-m] \tag{1.12}$$

and recalling from Eq. (1.9) that

$$y[n] = H(e^{j\omega})e^{j\omega n}$$

and substituting in Eq. (1.12), we get

$$H(e^{j\omega})e^{j\omega n} + \sum_{k=1}^{K} a_k H(e^{j\omega})e^{j\omega(n-k)} = \sum_{m=0}^{M} b_m e^{j\omega(n-m)}$$

which reduces to

$$H(e^{j\omega}) = \frac{\sum_{m=0}^{M} b_m e^{-j\omega m}}{1 + \sum_{k=1}^{K} a_k e^{-j\omega k}} \tag{1.13}$$

Example 1.10. Determine the frequency response of a certain LTI system that is defined by the following difference equation, then compute and plot the magnitude and phase response for $\omega = 0$ to 2π.

$$y[n] = x[n] + x[n-2] + 0.9y[n-1]$$

We first rewrite the equation to have all y terms on the left and all x terms on the right:

$$y[n] - 0.9y[n-1] = x[n] + x[n-2]$$

Then we get

$$H(e^{j\omega}) = \frac{1 + e^{-j\omega 2}}{1 - 0.9e^{-j\omega}}$$

which can be numerically evaluated using the following code:

```
N=10^3; dw=2*pi/N; w = 0:dw:2*pi-dw;
H = (1+exp(-j*2*w))./(1-0.9*exp(-j*w));
figure; subplot(2,1,1); plot(abs(H)); subplot(2,1,2); plot(angle(H))
```

1.8   REFERENCES

[1] William L. Briggs and Van Emden Henson, *The DFT, An Owner's Manual for the Discrete Fourier Transform*, SIAM, Philadelphia, 1995.

[2] Steven W. Smith, *The Scientist and Engineer's Guide to Digital Signal Processing*, California Technical Publishing, San Diego, 1997.

[3] Alan V. Oppenheim and Ronald W. Schaefer, *Discrete-Time Signal Processing*, Prentice-Hall, Englewood Cliffs, New Jersey, 1989.

[4] John G. Proakis and Dimitris G. Manolakis, *Digital Signal Processing, Principles, Algorithms, and Applications, Third Edition*, Prentice Hall, Upper Saddle River, New Jersey, 1996.

[5] Vinay K. Ingle and John G. Proakis, *Digital Signal Processing Using MATLAB V.4*, PWS Publishing Company, Boston, 1997.

1.9   EXERCISES

1. Determine analytically the DTFT of the following sequences:

    (a) $0.95^n u[n] - 0.8^{n-1} u[n-1]$

    (b) $\cos(2\pi (0:1:3)/4)$

(c) $(0.85)^{n-2}u[n-2]$

(d) $([0,0,(0.85)^{n-2}u[n-2]])u[n]$

(e) $(-0.9)^n u[n] + (0.7)^{n-3}u[n-3]$

(f) $n(u[n] - u[n-3])$

2. Compute the DTFTs of the following sequences and plot the magnitude and phase responses (use one of the scripts developed or presented earlier in this chapter, such as *LV_DTFT_Basic*). Be sure to increase the number of DTFT samples as the signal length increases. It's a good idea to use at least 10 times as many DTFT samples as the sequence length being evaluated. Consider plotting the magnitude on a logarithmic scale to see fine detail better. To avoid the problem of taking the logarithm of zero, add a small number such as 10^{-10} to the absolute value of the DTFT, then use the function *log10* and multiply the result by 20.

(a) **cos(2*pi*k*(0:1:N-1)/N)** for N = 8, 32, 128 and k = 0, 1, and $N/2$.

(b) **[ones(1,10),zeros(1,20),ones(1,10),zeros(1,20)]**

(c) **b = fir1(21,0.5)**

(d) **[1,0,1]**

(e) **[1,0,1,0,1,0,1]**

(f) **[1,0,1,0,1,0,1,0,1,0,1,0,1]**

(g) **[1,0,1,0,1,0,1,0,1,0,1].*hamming(13)'**

(h) **[1,0,1,0,1,0,1,0,1,0,1].*blackman(13)'**

(i) **[1,0,1,0,1,0,1,0,1,0,1].*kaiser(13,5)'**

(j) **[1,0,-1,0,1,0,-1,0,1,0,-1,0,1].*hamming(13)'**

(k) **[1,0,-1,0,1,0,-1,0,1,0,-1,0,1].*blackman(13)'**

(l) **[1,0,-1,0,1,0,-1,0,1,0,-1,0,1].*kaiser(13,5)'**

(m) **[real(j.^(0:1:10))].*blackman(11)'**

(n) **[real(j.^(0:1:20))].*blackman(21)'**

(o) **[real(j.^(0:1:40))].*blackman(41)'**

(p) **[real(j.^(0:1:80))].*blackman(81)'**

3. Write a script that can receive b and a difference equation coefficients (according to Eq. (1.13)) in row vector form (normalized so a_0 = 1), compute, and display the following:

(a) The unit impulse response of the system defined by b and a.

(b) The unit step response of the system.

(c) The magnitude and phase of the DTFT.

(d) The response to a linear chirp of length 1024 samples and frequencies from 0 to 512 Hz (0 to π radians in normalized frequency).

(e) The magnitude of response to a complex linear chirp of length 1024 samples and frequencies from 0 to 512 Hz (0 to π radians in normalized frequency). Such a chirp can be generated by the following code:

N = 1024; t = 0:1/N:1; y = chirp(t,0,1,N/2) + ...
j*chirp(t,0,1,N/2,'linear',90)

(f) The response to a signal of length 1024 samples containing a cosine of frequency 128.

(g) The response to a signal of length 1024 samples containing a cosine of frequency 256.

Use the script to evaluate the LTI systems defined by the following difference equations or b and a coefficients. You should note that some of the systems are not stable. Compare the plot of DTFT magnitude to the two chirp responses for each of the difference equations below. Where can the steady-state magnitude of the responses specified in (b), (f), and (g) above be found on the DTFT magnitude plot? State whether each of the systems is stable or unstable.

(I) $y[n] = x[n] + x[n-1] + 0.9y[n-1]$

(II) $y[n] = x[n] + 1.4y[n-1] - 0.81y[n-2]$

(III) $y[n] = x[n] + 1.4y[n-1] + 0.81y[n-2]$

(IV) $y[n] = x[n] - 2.45y[n-1] +2.37y[n-2] -0.945y[n-3]$

(V) $y[n] = x[n] +1.05y[n-1]$

(VI) $y[n] = 0.094x[n] + 0.3759x[n-1] + 0.5639x[n-2] + 0.3759x[n-3] + ...$
        $0.094x[n-4] + 0.486y[n-2] + 0.0177y[n-4]$

(VII) b = [0.6066,0,2.4264,0,3.6396,0,2.4264,0,0.6066];
        a = [1,0,3.1004,0,3.7156,0,2.0314,0,0.4332]

4. For the following different functions, compute the DTFT, then plot the function on a first subplot and its DTFT magnitude on a second subplot. The $sinc$ function can be evaluated using the function $sinc(x)$. What kind of filter impulse response results from the functions below? What effect does the parameter c have, and what effect does the length of the vector n have?

(a) **n =-9:1:9; c = 0.8; y = sinc(c*n);**

(b) **n =-9:1:9; c = 0.4; y = sinc(c*n);**

(c) **n =-9:1:9; c = 0.2; y = sinc(c*n);**

(d) **n =-9:1:9; c = 0.1; y = sinc(c*n);**

(e) **n =-39:1:39; c = 0.8; y = sinc(c*n);**

(f) **n =-39:1:39; c = 0.4; y = sinc(c*n);**

(g) **n =-39:1:39; c = 0.2; y = sinc(c*n);**

(h) **n =-39:1:39; c = 0.1; y = sinc(c*n);**

(i) **n =-139:1:139; c = 0.8.; y = sinc(c*n);**

(j) **n =-139:1:139; c = 0.4.; y = sinc(c*n);**

(k) **n =-139:1:139; c = 0.2.; y = sinc(c*n);**

(l) **n =-139:1:139; c = 0.1.; y = sinc(c*n);**

5. Compute and plot the magnitude and phase of the DTFT of the following impulse responses x. Note the effect of the frequency parameter f and the difference in impulse responses and especially phase responses between (a)-(e) and (f)-(j).

(a) **n = -10:1:10; f = 0.05; x = ((0.9).^(abs(n))).*cos(f*pi*n)**

(b) **n = -10:1:10; f = 0.1; x = ((0.9).^(abs(n))).*cos(f*pi*n)**

(c) **n = -10:1:10; f = 0.2; x = ((0.9).^(abs(n))).*cos(f*pi*n)**

(d) **n = -10:1:10; f = 0.4; x = ((0.9).^(abs(n))).*cos(f*pi*n)**

(e) n = -10:1:10; f = 0.8; x = ((0.9).^(abs(n))).*cos(f*pi*n)

(f) n = 0:1:20; f = 0.05; x = ((0.9).^(abs(n))).*cos(f*pi*n)

(g) n = 0:1:20; f = 0.1; x = ((0.9).^(abs(n))).*cos(f*pi*n)

(h) n = 0:1:20; f = 0.2; x = ((0.9).^(abs(n))).*cos(f*pi*n)

(i) n = 0:1:20; f = 0.4; x = ((0.9).^(abs(n))).*cos(f*pi*n)

(j) n = 0:1:20; f = 0.8; x = ((0.9).^(abs(n))).*cos(f*pi*n)

6. An ideal lowpass filter should have unity gain for $|\omega| \leq \omega_c$ and zero gain for $|\omega| > \omega_c$, and a linear phase factor or constant sample delay equal to $e^{-j\omega M}$ where M represents the number of samples of delay. Use the inverse DTFT to determine the impulse response that corresponds to this frequency specification. That is to say, perform the following integration

$$x[n] = \frac{1}{2\pi} \int_{-\pi}^{\pi} X(e^{j\omega}) e^{j wn} d\omega$$

where

$$X(e^{j\omega}) = \begin{cases} 1 \cdot e^{-j\omega M} & |\omega| \leq \omega_c \\ 0 & |\omega| > \omega_c \end{cases}$$

After performing the integration and obtaining an expression for $x[n]$, use n = -20:1:20 and M = 10, and obtain four impulse responses corresponding to the following values of ω_c:

(a) 0.1π

(b) 0.25π

(c) 0.5π

(d) 0.75π

Convolve each of the four resulting impulse responses with a chirp of length 1000 samples and frequency varying from 0 to 500 Hz. For each of the four impulse responses, plot the impulse response on one subplot and the chirp response on a second subplot. On a third subplot, plot the result from numerically computing the DTFT of each impulse response, or instead use the script *LV_DTFT_Basic* or the similar short version presented in the text to evaluate the DTFT and plot it in a separate window. To display on the third subplot, create a script

$$d = LVxDTFT\_Basic(x, M, R)$$

based on *LV_DTFT_Basic*, but which delivers the DTFT as the output argument d and which does not itself create a display.

7. For each of the four impulse responses computed above, obtain a new impulse response by subtracting the given impulse response from the vector [zeros(1,20),1,zeros(1,20)] (this assumes that the vector n in the previous example ran from -20 to +20).

Determine what kind of filter the four new impulse responses form. To do this, plot the magnitude of the DTFT of each original impulse response next to the DTFT of each corresponding new impulse response.

8. Create a script in accordance with the following call syntax, which should create plots similar to that shown in Fig. 1.3; test it with the sample calls given below.

function LVxDTFT(x,n,M,R,FreqOpt,FigNo)
% **Computes and displays the magnitude, phase, real, and**
% **imaginary parts of the DTFT of the sequence x having**
% **time indices n, evaluated over M samples.**
% **Pass R as 1 to evaluate over pi radians, or 2 to evaluate**
% **over 2*pi radians**
% **Use FreqOpt = 1 for symmetrical frequency evaluation**
% **about frequency 0 or FreqOpt = 2 for an asymmetrical**
% **frequency evaluation**
% **Sample calls:**
% **LVxDTFT([cos(2*pi*25*(0:1:99)/100)],[0:1:99],500,2,1,88)**
% **LVxDTFT([cos(2*pi*25*(-50:1:50)/100)],[-50:1:50],500,2,1,88)**
% **LVxDTFT([cos(2*pi*5*(0:1:20)/20)],[0:1:20],100,2,1,88)**
% **LVxDTFT([cos(2*pi*5*(-10:1:10)/20)],[-10:1:10],100,2,1,88)**
% **LVxDTFT([cos(2*pi*25*(0:1:100)/100)],[-50:1:50],500,2,1,88)**
% **LVxDTFT([exp(j*2*pi*25*(0:1:99)/100)],[0:1:99],500,2,1,88)**
% **LVxDTFT([cos(2*pi*25*(0:1:99)/100)],[0:1:99],1000,2,1,88)**

9. Create a script that conforms to the following call syntax, which should create plots similar to that shown in Fig. 1.6; test it with the sample calls given below.

function LVxDTFT_MS(x,SampOffset,FreqOffsetExp,...
% **M,R,TimeOpt,FreqOpt)**
% **Computes and displays the magnitude, phase, real, and**
% **imaginary parts of the DTFT of the sequence x, evaluated**
% **over M samples, then computes the same for a**
% **modified version of x that has been shifted by**
% **SampOffset samples and multiplied by a complex**
% **exponential FreqOffsetExp.**
% **Pass R as 1 to evaluate from 0 to pi radians, or**
% **Pass R as 2 to evaluate from 0 to 2*pi radians**
% **Pass TimeOpt as 1 to let n (the time indices for x1**
% **and x2) be computed as n = -(N-1)/2:1:(N-1)/2 for N odd**
% **or n = -N/2+1:1:N/2; for even N.**
% **Pass FreqOpt as 1 for symmetrical frequency computation**
% **and display (-R*pi to +R*pi, for example) or as 2 for**
% **frequency comp.and display from 0 to R*pi**
% **Sample calls:**
% **LVxDTFT_MS([cos(2*pi*25*(0:1:100)/100)],0,...**

```
% exp(j*2*pi*10*(0:1:100)/100),500,2,1,1)
% LVxDTFT_MS([cos(2*pi*25*(0:1:100)/100)],0,...
% exp(-j*2*pi*10*(0:1:100)/100),500,2,1,1)
% LVxDTFT_MS([cos(2*pi*25*(0:1:100)/100)],0,...
% exp(-j*pi/2),500,2,1,1) % shifts phase
% LVxDTFT_MS([exp(j*2*pi*25*(0:1:100)/100)],0,...
% exp(-j*2*pi*10*(0:1:100)/100),500,2,1,1)
% LVxDTFT_MS([cos(2*pi*25*(0:1:100)/100)],0,...
% exp(j*2*pi*12.5*(0:1:100)/100),1000,2,1,1)
% LVxDTFT_MS([1 0 1],2,1,300,2,1,1)
% LVxDTFT_MS([1 0 1],0,exp(j*2*pi*1*(0:1:2)/3),...
% 100,2,1,1)
% LVxDTFT_MS([cos(2*pi*25*(0:1:100)/100)],0,...
% exp(j*2*pi*12.5*(0:1:100)/100),200,2,1,1)
% LVxDTFT_MS([1 0 1],0,exp(j*2*pi*(0:1:2)/3),300,2,1,1)
```

10. Compute 1200 evenly distributed frequency samples of the DTFT (i.e., a numerical approximation of the DTFT) for the three following sequences:

   (a) [0.1,0.7,1,0.7,0.1]
   (b) [1,0,0,0,1]
   (c) $[0.95^n - 0.85^n]$ for $0 \leq n \leq \infty$

11. Numerically compute the inverse DTFTs of the three numeric DTFTs computed in the previous exercise (for item (c), compute the inverse DTFT for $n = 0:1:2$).

12. Compute the net output signal obtained by convolving the signal

$$x = 0.5\cos(2\pi n/16) + 0.3\sin(5\pi n/16)$$

with the impulse response [1,0,-1] two ways, first, by performing the time domain convolution, and second, by computing the DTFT of the impulse response to determine the magnitude and phase responses for the frequencies in the signal, and then scaling and shifting the two signal components, and finally summing the two scaled and shifted signal components to obtain the net output response. Verify that the results are the same during steady state.

CHAPTER 2

The z-Transform

2.1 OVERVIEW

In the previous chapter, we took a brief look at the Fourier and Laplace families of transforms, and a more detailed look at the DTFT, which is a member of the Fourier family which receives a discrete time sequence as input and produces an expression for the continuous frequency response of the discrete time sequence. With this chapter, we take up the z-transform, which uses correlators having magnitudes which can grow, decay, or remain constant over time. It may be characterized as a discrete-time variant of the Laplace Transform. The z-transform can not only be used to determine the frequency response of an LTI system (i.e., the LTI system's response to unity-amplitude correlators), it reveals the locations of poles and zeros of the system's transfer function, information which is essential to characterize and understand such systems. The z-transform is an indispensable transform in the discrete signal processing toolbox, and is virtually omnipresent in DSP literature. Thus, it is essential that the reader gain a good understanding of it.

The z-transform mathematically characterizes the relationship between the input and output sequences of an LTI system using the generalized complex variable z, which, as we have already seen, can be used to represent signals in the form of complex exponentials. Many benefits accrue from this:

- An LTI system is conveniently and compactly represented by an algebraic expression in the variable z; this expression, in general, takes the form of the ratio of two polynomials, the numerator representing the FIR portion of the LTI system, and the denominator representing the IIR portion.

- Values of z having magnitude 1.0, which are said to "lie on the unit circle" can be used to evaluate the z-transform and provide a frequency response equivalent to the DTFT.

- Useful information about a digital system can be deduced from its z-transform, such as location of system poles and zeros.

- Difference equations representing the LTI system can be constructed directly from inspection of the z-transform.

- An LTI system's impulse response can be obtained by use of the Inverse z-transform, or by constructing a digital filter or difference equation directly from the z-transform, and processing a unit impulse.

- The z-transform of an LTI system has, in general, properties similar or analogous to various other frequency domain transforms such the DFT, Laplace Transform, etc.

By the end of this chapter, the reader will have gained a practical knowledge of the z-transform, and should be able to navigate among difference equations, direct-form, cascade, and parallel filter topologies, and the z-transform in polynomial or factored form, converting any one representation to another. Additionally, an understanding will have been acquired of the inverse z-transform, and use of the z-transform to evaluate frequency response of various LTI systems such as the FIR and the IIR.

2.2   SOFTWARE FOR USE WITH THIS BOOK

The software files needed for use with this book (consisting of m-code (.m) files, VI files (.vi), and related support files) are available for download from the following website:

http://www.morganclaypool.com/page/isen

The entire software package should be stored in a single folder on the user's computer, and the full file name of the folder must be placed on the MATLAB or LabVIEW search path in accordance with the instructions provided by the respective software vendor (in case you have encountered this notice before, which is repeated for convenience in each chapter of the book, the software download only needs to be done once, as files for the entire series of four volumes are all contained in the one downloadable folder).

See Appendix A for more information.

2.3   DEFINITION & PROPERTIES

2.3.1   THE Z-TRANSFORM

The z-transform of a sequence $x[n]$ is:

$$X(z) = \sum_{n=-\infty}^{\infty} x[n]z^{-n}$$

where z represents a complex number. The transform does not converge for all values of z; the region of the complex plane in which the transform converges is called the **Region of Convergence (ROC)**, and is discussed below in detail. The sequence z^{-n} is a complex correlator generated as a power sequence of the complex number z and thus $X(z)$ is the correlation (CZL) between the signal $x[n]$ and a complex exponential the normalized frequency and magnitude variation over time of which are determined by the angle and magnitude of z.

2.3.2   THE INVERSE Z-TRANSFORM

The formal definition of the inverse z-transform is

$$x[n] = \frac{1}{2\pi j} \oint X(z)z^{n-1}dz \qquad (2.1)$$

where the contour of integration is a closed counterclockwise path in the complex plane that surrounds the origin ($z = 0$) and lies in the ROC.

There are actually many methods of converting a z-transform expression into a time domain expression or sequence. These methods, including the use of Eq. (2.1), will be explored later in the chapter.

2.3.3   CONVERGENCE CRITERIA

Infinite Length Causal (Positive-time) Sequence

When $x[n]$ is infinite in length, and identically 0 for $n < 0$, the ratio of $x[n]$ to z^n (or in other words, $x[n]z^{-n}$) must generally decrease in magnitude geometrically as n increases for convergence to a finite sum to occur.

Note that if the sequence $x[n]$ is a geometrically convergent series, then the z-transform will also converge provided that

$$\left| \frac{x[n+1]}{x[n]} \right| < |z|$$

In terms of numbers, if

$$\left| \frac{x[n+1]}{x[n]} \right| = 0.9$$

for example, then it is required that $|z| > 0.9$ for convergence to occur.

If in fact $x[n]$ is a geometrically convergent series, and z is properly chosen, the sum of the infinite series of numbers consisting of $x[n]z^{-n}$ may conveniently be written in a simple algebraically closed form.

Example 2.1.   Determine the z-transform for a single pole IIR with a real pole p having a magnitude less than 1.0.

The impulse response of such a filter may be written as

$$[p^0 (= 1), p^1, p^2, p^3, \dots p^n]$$

etc., or to pick a concrete example with the pole at 0.9,

$$0.9^n = [1, 0.9, 0.81, 0.729, \dots]$$

and the z-transform would therefore be:

$$A(z) = 1 + 0.9z^{-1} + 0.81z^{-2} + 0.729z^{-3} + \dots p^n z^{-n}$$

or in generic terms as

$$A(z) = 1 + pz^{-1} + p^2 z^{-2} + p^3 z^{-3} + \dots p^n z^{-n}$$

The summation of an infinite number of terms of the form c^n where $|c| < 1$ with

$$0 \leq n < \infty$$

is

$$\frac{1}{1 - c} \tag{2.2}$$

For the single pole IIR with a pole at p, and by letting $c = pz^{-1}$ (note that $p^0 = 1$) in Eq. (2.2), we get the closed-form z-transform as

$$A(z) = \frac{1}{1 - pz^{-1}} = \frac{z}{z - p} \tag{2.3}$$

The z-transform in this case is defined or has a finite value for all z with $|z| > |p|$ or

$$\left| \frac{p}{z} \right| < 1$$

Example 2.2.   Plot the ROC in the z-plane for the z-transform corresponding to the sequence $0.8^n u[n]$.

The z-transform is

$$X(z) = \frac{1}{1 - 0.8z^{-1}}$$

which converges for $|z| > 0.8$. The shaded area of Fig. 2.1 shows the ROC.

Example 2.3.   Write the z-transform for the sequence $0.7^n u[n]$, evaluate it at 500 values of z lying on the unit circle (i.e., having a magnitude of 1.0, and radian frequencies between 0 and 2π), and plot the magnitude of the result.

Values of z on the unit circle have magnitude greater than 0.7 and thus lie in the ROC for the z-transform, which is $1/(1- 0.7z^{-1})$; Figure 2.2 shows the result of running the following m-code, which computes and plots the magnitude of the z-transform:

```
radFreq = [0:2*pi/499:2*pi]; z = exp(j*radFreq);
Zxform = 1./(1-0.7*z.^(-1)); plot(radFreq/pi,abs(Zxform))
```

For the positive-time or causal sequence $x[n]$, convergence of the z-transform is guaranteed for all z with magnitude greater than the pole having greatest magnitude (sometimes referred to as the dominant pole) in a transfer function.

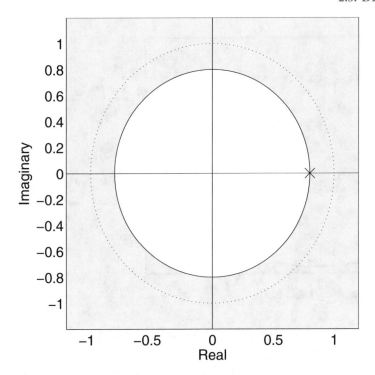

Figure 2.1: The Region of Convergence for a system having one pole at 0.8 is shown as the shaded area outside (not including) a circle of radius 0.8. The unit circle is shown as a dotted circle at radius 1.0.

For example, a transfer function having two poles at 0.9 and 0.8, respectively, would have the z-transform

$$A(z) = \frac{1}{(1 - 0.9z^{-1})(1 - 0.8z^{-1})}$$

which would converge for all z having $|z| > 0.9$.

The z-transform is undefined where it does not converge. Consider an example where $p = 0.9$, and we choose to evaluate the transform at $z = 0.8$. We write the first few terms of the definition of the z-transform as

$$A(z) = p^0 z^0 + pz^{-1} + p^2 z^{-2} + p^3 z^{-3} + \ldots p^n z^{-n}$$

which in concrete terms would be

$$A(z) = (\frac{0.9}{0.8})^0 + (\frac{0.9}{0.8})^1 + (\frac{0.81}{0.64})^2 + (\frac{0.9}{0.8})^3 + \ldots (\frac{0.9}{0.8})^n$$

which clearly diverges, and thus the transform is undefined.

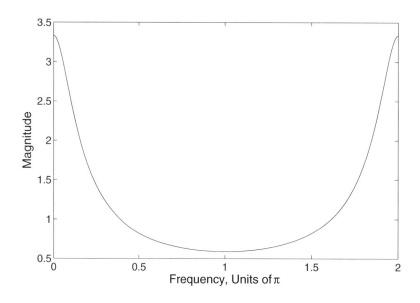

Figure 2.2: The magnitude of $H(z) = 1/(1 - 0.7z^{-1})$ evaluated at $z = \exp(j\omega)$ for $\omega = 0$ to 2π.

Note that the general formula for the z-transform of a single pole transfer function given in Eq. (2.3), if filled in with a value of z which is not in the ROC, will give an answer, but it will not be valid. Thus, it is very important to know what the region of convergence is for a given problem.

Example 2.4.   Determine, and verify using a numerical computation, the z-transform and ROC of the sequence defined as

$$x[n] = 0.9^n u[n - 1]$$

Taking into account the delay of one sample induced by the delayed unit step function, the z-transform would be

$$X(z) = \sum_{n=1}^{\infty} x[n]z^{-n} \tag{2.4}$$

the first few terms of which are $0.9z^{-1}, 0.81z^{-2}, 0.9^3z^{-3}$, etc., which is a geometric series having its first term FT equal to $0.9z^{-1}$ and the ratio R between successive terms equal to $0.9z^{-1}$. The sum of a geometric series is $FT/(1 - R)$, which leads to the z-transform as

$$X(z) = \frac{0.9z^{-1}}{1 - 0.9z^{-1}} \tag{2.5}$$

with ROC: $|z| > 0.9$.

To verify the validity of this expression, we can first compute the summation in Eq. (2.4) to a large number of samples until convergence is (essentially) reached, using a value of z lying in the ROC. Then we will substitute the chosen value of z in Eq. (2.5) and compare results, using the following code. Note that in accordance with Eq. (2.4), we start each sequence, $x[n]$ and the negative power sequence of z, at power 1 rather 0.

N = 150; z = 1; zpowseq = z.ˆ(-1:-1:-N); x = 0.9.ˆ(1:1:N);
NumAns = sum(zpowseq.*x)
zXformAns = (0.9/z)/(1-0.9*(1/z))

Note that as $|z|$ gets closer to 0.9, a larger value of N is necessary for convergence. For example, if $z = 0.91$, N will need to be about 1500.

Infinite Length, Negative Time Sequence
An infinite-length sequence that is identically zero for values of n equal to or greater than zero is called a negative-time sequence. A geometric sequence defined as

$$x[n] = -b^n u[-n-1]$$

where c has a finite magnitude and n runs from -1 to $-\infty$ has the z-transform

$$X(z) = -\sum_{n=-\infty}^{-1} b^n z^{-n} = -\sum_{n=-\infty}^{-1} (\frac{b}{z})^n = -\sum_{n=1}^{\infty} (\frac{z}{b})^n$$

which can be reduced to

$$X(z) = 1 - \sum_{n=0}^{\infty} (\frac{z}{b})^n = 1 - \frac{1}{1-z/b} = \frac{z}{z-b} = \frac{1}{1-bz^{-1}}$$

which converges when $|z| < |b|$.

Example 2.5.   A certain negative-time sequence is defined by $x[n] = -0.9^n u[-n-1]$ where $n = -1 : -1 : -\infty$. Determine the z-transform and the ROC. Pick a value of z in the ROC, numerically compute the z-transform, and then compare the answer to that obtained by substituting the chosen value of z into the determined z-transform.

The z-transform is $1/(1-0.9z^{-1})$, and the ROC includes all z having magnitude less than 0.9. The value of the transform for $z = 0.6$ is $1/(1-0.9/0.6) = -2.0$.

We can numerically compute the z-transform, say, for 30 points, using $z = 0.6$, which gives the answer as -2.0:

N = 30; b = 0.9; n = -N:1:-1; z = 0.6; sig = -(b.ˆn); z2n = z.ˆ(-n);
subplot(2,1,1); stem(n,sig); xlabel('n'); ylabel('Signal Amp')

subplot(2,1,2); stem(n,z2n); xlabel('n'); ylabel('z^ {-n}')
zXform = -sum((b.^n).*(z.^(-n)))

Infinite Length, Two-Sided

Sequences that have infinite extent toward both $+\infty$ and $-\infty$ are called two-sided and the ROC is the intersection of the ROC for the positive and negative sequences.

Example 2.6.   Determine the z-transform and ROC for the following two-sided sequence:

$$x[n] = 0.8^n u[n] - 0.9^n u[-n-1]$$

The net z-transform is the sum of the z-transforms for each of the two terms, and the ROC is the intersection of the ROCs for each. Thus, we get

$$X(z) = \frac{z}{z - 0.8} - \frac{z}{z - 0.9}$$

with the ROC for the left-hand term (the right-handed or positive-time sequence) comprising $|z| > 0.8$ and for the left-handed or negative-time sequence, the ROC comprises $|z| < 0.9$. The net ROC for the two-sided sequence is, therefore, the open annulus (ring) defined by

$$0.8 < |z| < 0.9$$

Note that it may happen that the two ROCs do not intersect, and thus the transform in such case is undefined, i.e., does not exist.

Example 2.7.   Determine the z-transform for the system defined by

$$x[n] = 0.9^n u[n] - 0.8^n u[-n-1]$$

Here we see that for the positive-time sequence, the ROC includes z having magnitude greater than 0.9, whereas for the negative-time sequence, z must have magnitude less than 0.8. In other words, there are no z that can satisfy both criteria, and thus the transform is undefined.

Note that the z-transform alone cannot uniquely define the underlying time domain sequence since, for example, the positive time sequence $0.9^n u[n]$ and the negative-time sequence $-0.9^n u[-n-1]$ have the same z-transform

$$\frac{1}{1 - 0.9z^{-1}}$$

Therefore, it is necessary to specify both the z-transform and the ROC to uniquely define the underlying time domain sequence.

Finite Length Sequence
When the sequence $x[n]$ is finite in length and $x[n]$ is bounded in magnitude, the ROC is generally the entire complex plane, possibly excluding z = 0 and/or z = ∞.

Example 2.8.   Evaluate the z-transform and state the ROC for the following three sequences:

(a) x[n] = [1,0,-1,0,1] with time indices [0,1,2,3,4].
(b) x[n] = [1,0,-1,0,1] with time indices [-5,-4,-3,-2,-1].
(c) x[n] = [1,0,-1,0,1] with time indices [-2,-1,0,1,2].

For (a) we get

$$z^0 - z^{-2} + z^{-4}$$

which has its ROC as the entire complex plane except for $z = 0$. For (b) we get

$$z^5 - z^3 + z^1$$

which has its ROC as the entire complex plane except for z = ∞. For (c) we get

$$z^2 - z^0 + z^{-2}$$

which has its ROC as the entire complex plane except for $z = 0$ and $z = \infty$.

2.3.4 SUMMARY OF ROC FACTS

- **Since convergence is determined by the magnitude of z, ROCs are bounded by circles.**

- **For finite sequences that are zero-valued for all $n < 0$, the ROC is the entire z-plane except for $z = 0$.**

- **For finite sequences that are zero-valued for all $n > 0$, the ROC is the entire z-plane except for $z = \infty$.**

- **For infinite length sequences that are causal (positive-time or right-handed), the ROC lies outside a circle having radius equal to the pole of largest magnitude.**

- **For infinite length sequences that are anti-causal (negative-time or left-handed), the ROC lies inside a circle having radius equal to the pole of smallest magnitude.**

- **For sequences that are composites of the above criteria, the sequence should be considered as the sum of a number of subsequences, and the ROC is generally the intersection of the ROCs of each subsequence.**

2.3.5   TRIVIAL POLES AND ZEROS

The generalized z-transform for an finite-length sequence may be written as

$$B(z) = b_0 + b_1 z^{-1} + b_2 z^{-2} + \dots + b_{N-1} z^{-(N-1)}$$

where the sequence length is N. If the right-hand side of this equation is multiplied by z^{N-1}/z^{N-1} (i.e., 1), the result is

$$B(z) = (b_0 z^{N-1} + b_1 z^{N-2} + b_2 z^{N-3} + \dots + b_{N-1})/z^{N-1}$$

which has $(N-1)$ poles at $z = 0$. These are referred to as **Trivial Poles**.

Likewise, the generalized expression for an IIR

$$A(z) = 1/(a_0 - a_1 z^{-1} - a_2 z^{-2} - \dots - a_{N-1} z^{-(N-1)})$$

if multiplied on the right by z^{N-1}/z^{N-1} will yield a transfer function with N -1 **Trivial Zeros** at $z = 0$.

The trivial poles of a finite sequence (or FIR) cause the z-transform to diverge at $z = 0$ as mentioned above.

2.3.6   BASIC PROPERTIES OF THE Z-TRANSFORM

Linearity

If the z-transforms of two functions $x_1[n]$ and $x_2[n]$ are $X_1(z)$ and $X_2(z)$, respectively, then the z-transform of

$$Z(c_1 x_1[n] + c_2 x_2[n]) = c_1 X_1(z) + c_2 X_2(z); \;\; \text{ROC: ROC}(x_1[n]) \cap \text{ROC}(x_2[n])$$

for all values of c_1 and c_2.

Shifting or Delay

If $X[z]$ is the z-transform of $x[n]$, then the z-transform of the delayed sequence $x[n - d]$ is $z^{-d} X[z]$ with the ROC the same as that for $x[n]$. This is a very useful property since it allows one to write the z-transform of difference equations by inspection.

Example 2.9.   Using the shifting property, write the z-transform of the following causal system:

$$y[n] = x[n] + ay[n - 1] + bx[n - 1]$$

The z-transform may be written as

$$Y(z) = X(z) + az^{-1} Y(z) + bz^{-1} X(z)$$

which simplifies to

$$\frac{Y(z)}{X(z)} = \frac{1 + bz^{-1}}{1 - az^{-1}}; \ \text{ROC}: |z| > |a|$$

Convolution

The z-transform of the convolution of two functions $x_1[n]$ and $x_2[n]$ is the product of the z-transforms of the individual functions. Stated mathematically,

$$Z(x_1[n] * x_2[n]) = X_1(z)X_2(z); \ \text{ROC}: \text{ROC}(x_1[n]) \cap \text{ROC}(x_2[n])$$

where the symbol $*$ is used here to mean convolution.

Example 2.10.   Compare the convolution of two sequences using linear convolution and the z-transform technique.

Consider the two time domain sequences [1,1,1] and [1,2,-1]. The time domain linear convolution is $y = [1,3,2,1,-1]$. Doing the problem using z-transforms, we get

$$Y(z) = (1 + z^{-1} + z^{-2})(1 + 2z^{-1} - z^{-2}) = [1 + 3z^{-1} + 2z^{-2} + z^{-3} - z^{-4}]$$

from which we can write the equivalent impulse response by inspection, [1,3,2,1,-1], which is identical to the time domain result.

Time Reversal or Folding

If

$$Z(x[n]) = X(z)$$

then

$$Z(x[-n]) = X(1/z)$$

and the ROC is inverted.

Example 2.11.   A sequence is $x[n] = [1, 0, -1, 2, 1]$, having time indices of $[2, 3, 4, 5, 6]$. Determine the z-transform of $x[-n]$.

The z-transform of $x[n]$ is

$$z^{-2} + 0z^{-3} - z^{-4} + 2z^{-5} + z^{-6}; \ \text{ROC}: |z| > 0$$

According to the folding property, the z-transform of $x[-n]$ should be

$$(1/z)^{-2} - (1/z)^{-4} + 2(1/z)^{-5} + (1/z)^{-6} = z^2 - z^4 + 2z^5 + z^6$$

The time reversed sequence is [1,2,-1,0,1] having time indices [-6,-5,-4,-3,-2], and computing the z-transform of $x[-n]$ directly we get

$$z^6 + 2z^5 - z^4 + 0z^3 + z^2; \text{ ROC: } |z| < \infty$$

Multiplication By a Ramp

$$Z(nx[n]) = -z\frac{dX(z)}{dz}; \text{ ROC: ROC}(x[n])$$

Example 2.12.   Determine the z-transform of the sequence $nu[n]$.

We get initially

$$X(z) = -z\frac{d}{dz}(\frac{1}{1 - z^{-1}}) = -z\frac{d}{dz}([1 - z^{-1}]^{-1})$$

$$= -z(-(1 - z^{-1})^{-2}(z^{-2})) = \frac{z^{-1}}{(1 - z^{-1})^2} = \frac{0z^0 + z^{-1}}{1 - 2z^{-1} + z^{-2}}$$

We can verify this using the following code, which yields the sequence $nu[n]$ i.e., [0,1,2...].

y = filter([0 1],[1,-2,1],[1,zeros(1,20)])

2.3.7   COMMON Z-TRANSFORMS

Sequence	z-Transform	ROC
$\delta[n]$	1	$\forall z$
$u[n]$	$1/(1 - z^{-1})$	$|z| > 1$
$-u[-n-1]$	$1/(1 - z^{-1})$	$|z| < 1$
$a^n u[n]$	$1/(1 - az^{-1})$	$|z| > |a|$
$-b^n u[-n-1]$	$1/(1 - bz^{-1})$	$|z| < |b|$
$[a^n \sin \omega_o n]u[n]$	$\frac{(a\sin\omega_o)z^{-1}}{1-(2a\cos\omega_o)z^{-1}+a^2z^{-2}}$	$|z| > |a|$
$[a^n \cos \omega_o n]u[n]$	$\frac{(a\cos\omega_o)z^{-1}}{1-(2a\cos\omega_o)z^{-1}+a^2z^{-2}}$	$|z| > |a|$
$na^n u[n]$	$\frac{az^{-1}}{(1-az^{-1})^2}$	$|z| > |a|$
$-nb^n u[-n-1]$	$\frac{bz^{-1}}{(1-bz^{-1})^2}$	$|z| < |b|$

Example 2.13.   Represent in the z-domain the time domain convolution of the sequence [1, -0.9] with a single-pole filter having its pole at 0.9. Use the representation to determine the time domain convolution.

The convolution in the time domain of the two sequences can be equivalently achieved in the z-domain by multiplying the z-transforms of each time domain sequence, i.e., in this case

$$\frac{(1 - 0.9z^{-1})}{(1 - 0.9z^{-1})} = 1$$

Looking at the table of common z-transforms, we see that the unit impulse is the time domain signal that has as its z-transform the value 1. This can be confirmed computationally (for the first 20 samples) using the following code:

imp = 0.9.ˆ(0:1:100); sig = [1,-0.9]; td = conv(imp,sig)

which returns the value of *td* as a unit impulse sequence. Note that even though the single pole generates an infinitely long impulse response, this particular input sequence results in a response which is identically zero after the first sample of output.

Example 2.14.   The sequence randn(1,8) is processed by an LTI system represented by the difference equation below; specify another difference equation which will, when fed the output of the first LTI system, return the original sequence.

$$y[n] = x[n] + x[n - 2] + 1.2y[n - 1] - 0.81y[n - 2]$$

We rewrite the difference equation as

$$y[n] - 1.2y[n - 1] + 0.81y[n - 2] = x[n] + x[n - 2]$$

and then write the z-transform as

$$Y(z)(1 - 1.2z^{-1} + 0.81z^{-2}) = X(z)(1 + z^{-2})$$

which yields

$$Y(z)/X(z) = \frac{1 + z^{-2}}{1 - 1.2z^{-1} + 0.81z^{-2}}$$

The *b* and *a* coefficients for this system are *b* = [1,1] and *a* = [1,-1.2,0.81]. To obtain the inverse system, simply exchange the values for *b* and *a*, i.e., obtain the reciprocal of the z-transform. To check the answer, we will first process the input sequence with the original difference equation, then process the result with the inverse difference equation:

```
x = randn(1,8), y = filter([1,1],[1,-1.2,0.81],x);
ans = filter([1,-1.2, 0.81],[1,1],y)
```

2.3.8 TRANSFER FUNCTIONS, POLES, AND ZEROS

LTI System Representation

If an LTI system is represented by the difference equation

$$y[n] + \sum_{k=1}^{N-1} a_k y[n-k] = \sum_{m=0}^{M-1} b_m x[n-m]$$

Taking the z-transform and using properties such as the shifting property, we get

$$Y(z)(1 + \sum_{k=1}^{N-1} a_k z^{-k}) = X(z) \sum_{m=0}^{M-1} b_m z^{-m}$$

which yields the system transfer function as

$$H(z) = \frac{\sum_{m=0}^{M-1} b_m z^{-m}}{1 + \sum_{k=1}^{N-1} a_k z^{-k}} \tag{2.6}$$

Example 2.15. Determine the system transfer function for the LTI system represented by the following difference equation:

$$y[n] = 0.2x[n] + x[n-1] + 0.2x[n-2] + 0.95y[n-1]$$

We rewrite the difference equation as

$$y[n] - 0.95y[n-1] = 0.2x[n] + x[n-1] + 0.2x[n-2]$$

and then take the z-transform (using the shifting property)

$$Y(z) - 0.95Y(z)z^{-1} = 0.2X(z) + X(z)z^{-1} + 0.2X(z)z^{-2}$$

which yields

$$Y(z)/X(z) = \frac{0.2 + z^{-1} + 0.2z^{-2}}{1 - 0.95z^{-1}}$$

Eq. (2.6) can be expressed as the ratio of products of individual zero and pole factors by converting it to an expression in positive powers of z:

$$H(z) = (\frac{z^{N-1}}{z^{M-1}})(\frac{z^{M-1}}{z^{N-1}}) \frac{b_0 + b_1 z^{-1} + ...b_{(M-1)}z^{-(M-1)}}{1 + a_1 z^{-1} + ...a_{(N-1)}z^{-(N-1)}}$$

which becomes

$$b_0 z^{N-M} \left(\frac{z^{M-1} + (b_1/b_0)z^{M-2} + \dots + (b_{(M-2)}/b_0)z^1 + (b_{(M-1)})/b_0}{z^{N-1} + a_1 z^{N-1} + \dots a_{(N-2)}z^1 + a_{(N-1)}} \right)$$

which can then be written as the ratio of the product of pole and zero factors:

$$H(z) = b_0 z^{N-M} \frac{\Pi_{m=1}^{M-1}(z - z_m)}{\Pi_{k=1}^{N-1}(z - p_k)} \tag{2.7}$$

where z_m and p_k are the zeros and poles, respectively. Each of the factors may be interpreted, for any specific value of z, as a complex number having a magnitude and angle, and thus we can write

$$H(z) = b_0 z^{N-M} \frac{\Pi_{m=1}^{M-1} M_m \angle \theta_m}{\Pi_{k=1}^{N-1} M_k \angle \theta_k} \tag{2.8}$$

(An example pertaining to Eq. (2.8) will be presented below).

FIR Zeros

The difference equation for an FIR of length M is

$$y[n] = b_0 x[n] + b_1 x[n-1] + b_2 x[n-2] + \dots b_{(M-1)} x[n-(M-1)]$$

and translates (using the time-shifting property) directly into the z-domain as follows:

$$Y(z) = b_0 X(z) + b_1 X(z) z^{-1} + b_2 X(z) z^{-2} + \dots b_{(M-1)} X(z) z^{-(M-1)}$$

Then

$$\frac{Y(z)}{X(z)} = b_0 + b_1 z^{-1} + b_2 z^{-2} + \dots b_{(M-1)} z^{-(M-1)} \tag{2.9}$$

The roots of Eq. (2.9) are the zeros of the transfer function, i.e., values of z such that

$$b_0 + b_1 z^{-1} + b_2 z^{-2} + \dots b_{(M-1)} z^{-(M-1)} = 0$$

While Eq. (2.9) is in the form of a polynomial in negative powers of z, Eq. (2.10) shows the factored form

$$\frac{Y(z)}{X(z)} = \Pi_{m=1}^{M-1}(1 - r_m z^{-1}) \tag{2.10}$$

where r_m refers to the mth root of Eq. (2.9).

Example 2.16.   Obtain the z-transform and its roots for the finite impulse response $[1, 0, 0, 0, 1]$.

A simple filter having a periodic or comb-like frequency response can be formed by adding a single delayed version of a signal with itself. The impulse response

$$[1, 0, 0, 0, 1]$$

has, by definition, the z-transform

$$Y(z)/X(z) = 1 + z^{-4} \tag{2.11}$$

with ROC being the entire z-plane less the origin (z = 0).

The roots of Eq. (2.11) can be found using DeMoivre's Theorem. Setting Eq. (2.11) equal to zero and multiplying both sides of the equation by z^4 we get

$$z^4 = -1$$

By expressing -1 as odd multiples of π radians (180 degrees) in the complex plane, we get

$$z^4 = 1\angle(\pi + 2\pi m) = 1\angle\pi(2m + 1)$$

where m can be 0, 1, 2, or 3. The roots are obtained, for each value of m, by taking the fourth root of the magnitude, which is 1, and one-quarter of the angle. We would thus get the four roots as $\pi/4$, $3\pi/4$, $5\pi/4$, and $7\pi/4$. Plots of the magnitude of the z-transform and the transfer function zeros are shown in Fig. 2.3. The z-plane plot shown in Fig. 2.3, plot (b), can be obtained using the function

$$zplane(z, p)$$

where z and p are column vectors of zeros and poles of the transfer function, or, alternately

$$zplane(b, a)$$

where b and a are row vectors of the z-transform numerator and denominator coefficients, respectively. The m-code used to generate Fig. 2.3 was

```
freq = -pi:0.02:pi; z = exp(j*freq);
FR = abs(1 + z.^(-4));
figure(44); subplot(211);
plot(freq/pi,FR)
xlabel('Frequency, Units of π')
ylabel('Magnitude')
subplot(212);
zplane([1 0 0 0 1],1)
xlabel('Real Part')
ylabel('Imaginary Part')
```

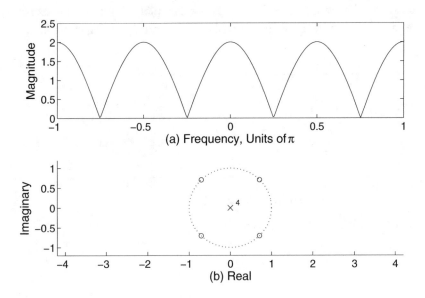

Figure 2.3: (a) Magnitude of the z-transform of a unity-additive comb filter having a delay of four samples, having the z-transform $H(z) = 1 + z^{-4}$; (b) Plot of the zeros of the z-transform of the comb filter of (a). Note the four poles at $z = 0$.

For more complicated transfer functions, the roots can be found using the function *roots*.

Example 2.17. Obtain the z-transform and its roots for the finite impulse response $[1, 0, 0, 0, 0.5]$.

The z-transform is

$$Y(z)/X(z) = 1 + 0.5z^{-4}$$

The zeros of this transfer function do not lie on the unit circle. They can be found as

$$z = \sqrt[4]{0.5}\angle((\pi + 2\pi m)/4)$$

where $m = 0,1,2,$ and 3, by making this m-code call

m = 0:1:3; angs = exp(j*(pi*(2*m +1)/4)); theZeros = 0.5 ^(0.25)*angs

or by making the call

theZeros = roots([1,0,0,0,0.5])

Figure 2.4, plot (b), shows the zeros returned by the above call, plotted in the complex plane, while plot (a) shows the magnitude of the frequency response of the corresponding comb filter.

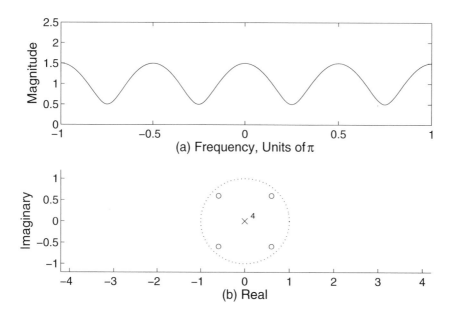

Figure 2.4: (a) Magnitude of the z-transform $H(z) = 1 + 0.5z^{-4}$ over the frequency range $-\pi$ to π radians; (b) Plot of the zeros of the z-transform of the comb filter of (a).

Example 2.18.   Determine the roots (zeros) of the transfer function corresponding to the impulse response $[ones(1, 6)]$.

The z-transform of the impulse response is

$$H(z) = 1 + z^{-1} + z^{-2} + z^{-3} + z^{-4} + z^{-5}$$

The transfer function length is $M + 1 = 6$, and therefore there are $M = 5$ roots, obtained by the call

$$\textbf{rts = roots([1,1,1,1,1,1])}$$

which yields the answer $0.5 \pm 0.866j$, -1, $-0.5 \pm 0.866j$.

IIR-Poles

The z-transform for a single pole IIR is

$$\frac{Y(z)}{X(z)} = \frac{1}{1 - pz^{-1}} \tag{2.12}$$

The value of p in Eq. (2.12) is called a pole, because if z assumes the same value, the transfer function becomes infinite or undefined, since

$$\lim_{z \to p} (1/(1 - p(\frac{1}{z}))) \to \infty$$

A multiple-pole transfer function can be written in factored form as

$$\frac{Y(z)}{X(z)} = 1/ \Pi_{n=1}^{N}(1 - p_n z^{-1})$$

where p_n are the poles and N is the total number of poles. The coefficient or polynomial form can be arrived at by performing the indicated multiplication.

Example 2.19.   A certain LTI system has poles at $\pm 0.6j$. Determine the z-transform in polynomial form.

We perform the multiplication

$$H(z) = \frac{1}{(1 - (j0.6)z^{-1})(1 - (-j0.6)z^{-1})} = \frac{1}{1 + 0.36z^{-2}}$$

This multiplication of polynomials can be performed as a convolution when a vector of roots is available:

> **theRoots = [j*0.6,-j*0.6];**
> **NetConv = [1];**
> **for ctr = 1:1:length(theRoots)**
> **NetConv = conv(NetConv,[1 -theRoots(ctr)]);**
> **end**
> **theCoeff = NetConv**

which returns

theCoeff = [1 ,0, 0.36]

Example 2.20.   Compute the z-transform coefficients of an IIR having poles at $0.9j$ and $-0.9j$; check your results using the MathScript function *poly*.

The coefficients are

conv([1, -(0 + 0.9j)],[1, -(0 - 0.9j)]) = [1,0,0.81]

and the z-transform is

$$Y(z)/X(z) = 1/(1 + 0z^{-1} + 0.81z^{-2})$$

This can be checked using the function *poly* which converts a vector of roots into polynomial coefficients:

$$a = poly([0.9^*j, -0.9^*j]) = [1, 0, 0.81]$$

Example 2.21.   The poles of a certain causal LTI system are at 0.9 and $\pm 0.9j$. There are three zeros at -1. Using Eq. (2.8), determine the magnitude of this system's response at DC. Use the function poly to obtain the polynomial form of the z-transform, and evaluate it to check your work.

We note that with three zeros, $M = 4$ and with three poles, $N = 4$. To get b_0, we must obtain the polynomial coefficients for the numerator of the z-transform. Note that the ROC is for $|z| > 0.9$.

zzs = [-1,-1,-1]; pls = [0.9, 0.9*j,-0.9*j]; Denom = poly(pls);
N = length(pls) +1; M = length(zzs) +1; Num = poly(zzs);
b0 = Num(1); z = 1; NProd = 1; for NCtr = 1:1:M-1;
NProd = NProd*(abs(z - zzs(NCtr))); end
DProd = 1; for DCtr = 1:1:N-1;
DProd = DProd*(abs(z - pls(DCtr))); end
MagResp = abs(b0)*abs(z^(N-M))*NProd/DProd
AltMag = abs(sum(Num.*(z.^(0:-1:-(M-1)))))/sum(Denom.* (z.^(0:-1:-(N-1)))))

2.3.9   POLE LOCATION AND STABILITY

The impulse response of a single pole IIR is $[1,p,p^2,p^3...]$. It can be seen that if $|p| < 1$, then a geometrically convergent series results. A bounded signal (one whose sample values are all finite) will not produce an unbounded output. If, however, $|p| > 1$ then the impulse response does not decay away, and a bounded input signal can produce an output that grows without bound. For the borderline case when $|p| = 1$, the unit impulse sequence as input produces a unit step ($u[n]$) as the corresponding output sequence. If the input signal is $u[n]$, then the output will grow without bound.

The difference equation for the single-pole IIR

$$y[n] = x[n] + py[n - 1]$$

can be used to observe the relationship between pole magnitude, stability, and pole location in the complex plane by choosing different values of p and then processing a test signal $x[n]$ such as the unit impulse $[1,0,0 ...]$ or the unit step $[1,1,1 ...]$. For $|p| \geq 1$ and the unit impulse as $x[n]$, the output $y[n]$ continues growing without further input.

For an LTI system to be stable, all of its poles must have magnitude less than 1.0.

Example 2.22.   Determine the pole magnitudes of the LTI system represented by the z-transform

$$H(z) = \frac{1}{1 - 1.9z^{-1} + 0.95z^{-2}}$$

We run the code

$$pmags = abs(roots([1,-1.9, 0.95]))$$

which yields *pmags* = [0.9746, 0.9746], meaning that the system is stable.

In summary

- **The poles of a transfer function must all have magnitudes less than 1.0 in order for the corresponding system to have a stable, bounded response to a bounded input signal (i.e., BIBO).**

- **Stable poles, when graphed in the z-domain (i.e., the complex plane), all lie inside the unit circle.**

Example 2.23.   For an IIR having a single real pole p, plot the first 45 samples of the impulse and unit step responses for the following values of p:

(a) 0.9
(b) 1.0
(c) 1.01

We can use this code to generate the impulse and step responses:

p = 0.9; ImpResp = filter(1,[1,-p],[1,zeros(1,44)])
StepResp = filter(1,[1,-p],[ones(1,95)])

Figure 2.5 shows the results for each of the three pole values. Note that for the one stable case (p = 0.9), the impulse response decays to zero, and the step response converges to a finite value.

You can experiment with pole location for single or complex conjugate pairs of poles and the corresponding/resultant system impulse response by calling the VI

DemoDragPolesImpRespVI.

Figure 2.6 shows an example of the VI *DemoDragPolesImpRespVI* using a pair of complex poles having an approximate magnitude of 0.9 and angles of approximately $\pm\pi/4$ radians.

A MATLAB script that performs a similar function is

ML_DragPoleZero

which, when called on the Command Line, opens up a GUI that allows you to select a single or complex conjugate pair of poles or zeros and move the cursor around the z-plane to select the pole(s)

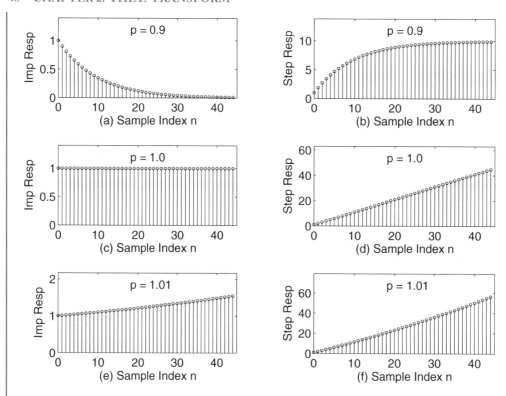

Figure 2.5: (a) Impulse response for a single pole IIR, p = 0.9; (b) Step response for same; (c) Impulse response for a single pole IIR, p = 1.0; (d) Step response for same; (e) Impulse response for a single pole IIR, p = 1.01; (f) Step response for same.

or zero(s). The magnitude and phase response of the z-transform and the real and imaginary parts of the impulse response are dynamically displayed as you move the cursor in the z-plane. For poles of magnitude 1.0 or greater, instead of using the z-transform to determine the frequency response (which is not possible since values of z along the unit circle are not in the ROC when the poles have magnitude 1.0 or greater), the DFT of a finite length of the test impulse response is computed. While this is not a true frequency response (none exists for systems having unstable poles since the DTFT does not converge), it gives an idea of how rapidly the system response to a unity-magnitude complex exponential grows as pole magnitude increases beyond 1.0.

2.4   CONVERSION FROM Z-DOMAIN TO TIME DOMAIN

There are a number of methods for computing the values of, or determining an algebraic expression for, the time domain sequence that underlies a given z-transform. Here we give a brief summary of some of these methods:

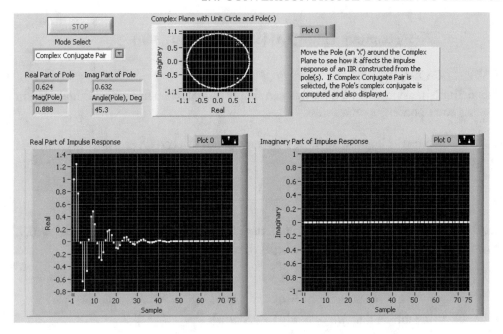

Figure 2.6: A VI allowing the user to drag a pole or pair of complex conjugate poles in the z-plane and observe the impulse response of an IIR constructed from the pole or poles. The first 75 samples of the impulse response are computed and displayed.

2.4.1 DIFFERENCE EQUATION

A simple, direct method is to write the difference equation of the system by inspection from the z-transform, and then process a unit impulse.

Example 2.24.   Compute the impulse response that corresponds to a causal LTI system that has the z-transform

$$X(z) = \frac{1 + z^{-1}}{1 - 0.9z^{-1}}$$

The difference equation is

$$y[n] = x[n] + x[n-1] + 0.9y[n-1]$$

with $y[n] = x[n] = 0$ for $n < 0$. We can compute an arbitrary number of samples N of the impulse response using the function *filter* with $b = [1,1]$, $a = [1, -0.9]$, and $x = [1, \text{zeros}(1,N)]$. Thus, we make the call

$$\text{N=50; ImpResp = filter([1,1],[1,-0.9],[1,zeros(1,50)])}$$

2.4.2   TABLE LOOKUP

In this method, the z-transform is in a form that can simply be looked up in a table of time domain sequences versus corresponding z-transforms.

Example 2.25.   Compute the impulse response that corresponds to a certain causal LTI system that has the z-transform

$$X(z) = \frac{1}{1 - 0.9z^{-1}}$$

Since the system is causal (i.e., positive-time), the ROC will include all z such that $|z| > 0.9$. We note that the z-transform of the sequence $a^n u[n]$ where $|a| < 1$ is

$$\frac{1}{1 - az^{-1}}$$

and hence are able to determine that $h[n] = a^n u[n] = 0.9^n u[n]$.

Example 2.26.   Compute the impulse response that corresponds to a certain causal LTI system that has the z-transform

$$X(z) = \frac{1 + z^{-1}}{1 - 0.9z^{-1}}$$

Since the system is causal (i.e., positive-time), the ROC will include all z such that $|z| > 0.9$. We can exercise a little ingenuity by breaking this z-transform into the product of the numerator and denominator, obtain the impulse response corresponding to each, and then convolve them in the time domain. We will thus convolve the sequences $0.9^n u[n]$ and $[1, 1]$. A suitable call to compute the first 30 samples is

$$\text{ImpResp = conv([0.9.ˆ(0:1:29)],[1,1])}$$

We can check this with the call

$$\text{ImpRespAlt = filter([1,1],[1,-0.9],[1,zeros(1,29)])}$$

2.4.3   PARTIAL FRACTION EXPANSION

In this method, the z-transform is not in a form that can readily be looked up. A z-transform of the general form

$$X(z) = \frac{b_0 + b_1 z^{-1} + ... + b_M z^{-M}}{1 + a_1 z^{-1} + ... + a_N z^{-N}} \tag{2.13}$$

can be rewritten as

$$X(z) = \frac{b_0' + b_1' z^{-1} + ... + b_{N-1}' z^{-(N-1)}}{1 + a_1 z^{-1} + ... + a_N z^{-N}} + \sum_{k=0}^{M-N} C_k z^{-k} \qquad (2.14)$$

which can be rewritten as a sum of fractions, one for each pole, plus a sum of polynomials if $M \geq N$.

$$X(z) = \sum_{n=1}^{N} \frac{R_n}{1 - p_n z^{-1}} + \sum_{k=0}^{M-N} C_k z^{-k} \qquad (2.15)$$

The values R_n are called residues, p_n is the n-th pole of $X(z)$, and it is assumed that the poles are all distinct, i.e., no duplicate poles. In such a case, the residues can be computed by:

$$R_n = \frac{b_0' + b_1' z^{-1} + ... + b_{N-1}' z^{-(N-1)}}{1 + a_1 z^{-1} + ... + a_N z^{-N}} (1 - p_n z^{-1}) \big|_{z=p_n}$$

The above expressions are true for distinct p_n. If p_n consists of S duplicate poles, the partial fraction expansion is

$$\sum_{s=1}^{S} \frac{R_{n,s} z^{-(s-1)}}{(1 - p_n z^{-1})^s} = \frac{R_{n,1}}{1 - p_n z^{-1}} + \frac{R_{n,2} z^{-1}}{(1 - p_n z^{-1})^2} + ... + \frac{R_{n,s} z^{-(S-1)}}{(1 - p_n z^{-1})^S}$$

The time domain impulse response can be written from the partial fraction expansion as

$$x[n] = \sum_{n=1}^{N} R_n Z^{-1} \left[\frac{1}{1 - p_n z^{-1}} \right] + \sum_{k=0}^{M-N} C_k \delta[n - k]$$

For causal sequences, the inverse z-transform of

$$\frac{1}{1 - p_n z^{-1}}$$

is

$$p_n^k u[k]$$

The function

$$[R, p, C_k] = residuez(b, a)$$

provides the vector of residues R, the corresponding vector of poles p, and, when the order of the numerator is equal to or greater than that of the denominator, the coefficients C_k of the polynomial in z^{-1}.

Example 2.27.   Construct the partial fraction expansion of the z-transform below; use the function residuez to obtain the values of R, p, and C_k. Determine the impulse response from the result.

$$X(z) = \frac{0.1 + 0.5z^{-1} + 0.1z^{-2}}{1 - 1.2z^{-1} + 0.81z^{-2}}$$

We make the call

[R,p,Ck] = residuez([0.1,0.5,0.1],[1,-1.2,0.81])

and receive results

R = [(-0.0117 -0.4726i),(-0.0117 + 0.4726i)]
p = [(0.6 + 0.6708i),(0.6 - 0.6708i)]
Ck = [0.1235]

from which we construct the expansion as

$$X(z) = \frac{R(1)}{1 - p(1)z^{-1}} + \frac{R(2)}{1 - p(2)z^{-1}} + Ck(1)z^0$$

and the impulse response is

$$x[n] = R(1)p(1)^n u[n] + R(2)p(2)^n u[n] + Ck(1)\delta[n] \tag{2.16}$$

A script that will numerically evaluate (2.16) and then compute the impulse response using *filter* with a unit impulse is

[R,p,Ck] = residuez([0.1,0.5,0.1],[1,-1.2,0.81])
n = 0:1:50; x = R(1)*p(1).^n + R(2)*p(2).^n; x(1) = x(1) + Ck;
altx = filter([0.1,0.5,0.1],[1,-1.2,0.81],[1 zeros(1,50)])
diff = x - altx, hold on; stem(x); stem(altx)

2.4.4   CONTOUR INTEGRATION IN THE COMPLEX PLANE
Numerical Method
For this method, we use the formal definition of the inverse z-transform:

$$x[n] = \frac{1}{2\pi j} \oint X(z)z^{n-1} dz \tag{2.17}$$

where the contour of integration is a closed counterclockwise path in the complex plane that surrounds the origin ($z = 0$) and lies in the ROC.

The integral at (2.17) can be evaluated directly using numerical integration. The script (see exercises below)

$$LVxNumInvZxform(NumCoefVec, DenCoefVec, M, ContourRad, nVals)$$

allows you to enter the z-transform of a causal sequence as its numerator and denominator coefficient vectors in ascending powers of z^{-1}. You also specify the number of points M along the contour to use to approximate the integral, the radius of the circle *ContourRad* that will serve as the contour, and $nVals$, the values of n (for $x[n]$) to be computed; $nVals$ must consist of nonnegative integers corresponding to the sample indices of a causal sequence $x[n]$.

The script computes the answer and generates two plots, the first showing the real part of $x[n]$ and the second showing the imaginary part of $x[n]$. If the z-transform coefficients are real, then $x[n]$ will be real, and the imaginary part of $x[n]$ should be zero. If this is the case, then it is possible to judge the accuracy of $x[n]$ (i.e., $real(x[n])$) by judging how close $imag(x[n])$ is to zero. Generally, the closer $imag(x[n])$ is to zero, the more accurate is $real(x[n])$.

A suitable call to obtain the inverse of the z-transform

$$X(z) = \frac{1}{1 - 1.2z^{-1} + 0.81z^{-2}} \quad \text{(ROC: } |z| > 0.9)$$

is

LVxNumInvZxform([1],[1,-1.2, 0.81],10000,1,[0:1:50])

which results in Fig. 2.7. Theoretically, any counterclockwise closed contour in the ROC will result in the same answer; circular contours are particularly easy to specify and compute. You can verify that any circular contour in the ROC will give the same answer by changing the value of *ContourRad*.

Example 2.28.   Estimate, using numerical contour integration, the time domain sequence corresponding to the z-transform

$$X(z) = \frac{z^{-1}}{1 - 2z^{-1} + z^{-2}} \quad \text{(ROC: } |z| > 1.0)$$

We make the call

LVxNumInvZxform([0,1],[1,-2,1],10000,1.05,[0:1:50])

which results in Fig. 2.8.

The call immediately above yields the sequence

$$x[n] = nu[n]$$

and thus you can see how accurate the result is since the answer should be a sequence of integers; the imaginary parts should all be zero-valued. To improve accuracy, increase M; to decrease computation time, at the expense of accuracy, decrease M.

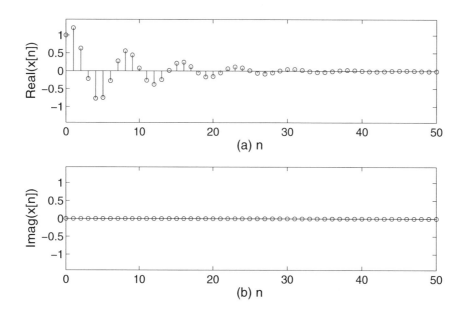

Figure 2.7: (a) The real part of the first 50 samples of the impulse response corresponding to the z-transform $1/(1 -1.2z^{-1}+0.81z^{-2})$, computed using numerical contour integration along a circular contour of radius 1.0 since the largest magnitude pole is 0.9; (b) Imaginary part of same.

Sum of Residues

The traditional way to evaluate (2.17) is to use a theorem (the Cauchy Residue Theorem) that states that the value of a contour integral in the z-plane over any closed counterclockwise path that encircles the origin and lies within the region of convergence is equal to the sum of the residues of the product of $X(z)$ and z^{n-1} within the chosen contour, which may be, for example, for right-handed z-transforms (corresponding to a causal sequence $x[n]$), a circle around the origin whose radius is greater than the magnitude of the largest pole of $X(z)$. For simple (nonrepeated) poles, all of which lie within the chosen contour, this statement may be stated mathematically as

$$x[n] = \frac{1}{2\pi j} \oint X(z)z^{n-1}dz = \sum_k (z - z_k)X(z)z^{n-1} \big|_{z=z_k}$$

Executing this method requires that the expression $X(z)z^{n-1}$ be expanded by partial fractions so that the term $(z - z_k)$ will cancel the corresponding pole in one fraction and reduce the others to zero when z_k is substituted for z.

Example 2.29. Using the sum of residues method, determine $x[n]$ for a certain causal LTI system that has poles at $\pm 0.9j$ and the corresponding z-transform

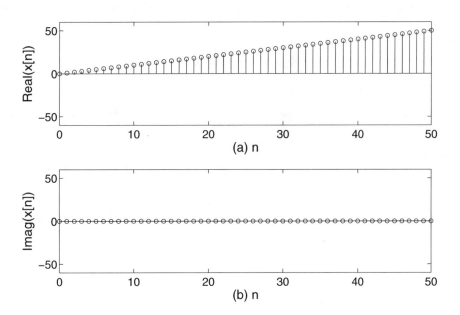

Figure 2.8: (a) The real part of the first 51 samples of the numerical approximation to the inverse z-transform of the transfer function $(z^{-1})/(1 - 2z^{-1} + z^{-2})$; (b) Imaginary part of same.

$$X(z) = \frac{1}{(1 - j0.9z^{-1})(1 + j0.9z^{-1})} = \frac{1}{1 + 0.81z^{-2}}$$

Making the call

$$[\mathbf{R,p,Ck}] = \mathbf{residuez([1],[1,0,0.81])}$$

allows us to write the partial fraction expansion of $X(z)$ (in positive powers of z) as

$$X(z) = \frac{0.5z}{(z - 0.9j)} + \frac{0.5z}{(z + 0.9j)}$$

and the net integrand as

$$\frac{0.5z^n}{(z - 0.9j)} + \frac{0.5z^n}{(z + 0.9j)}$$

We obtain the first residue by multiplying the expression by the first pole $(z - 0.9j)$ and then setting $z = 0.9j$ from which we get

$$0.5(0.9j)^n + \frac{0.5(0.9j)^n(0.9j - 0.9j)}{(0.9j + 0.9j)} = 0.5(0.9j)^n$$

Doing the same thing with the second pole, $-0.9j$, we get $0.5(-0.9j)^n$. Adding the two residues together we get

$$x[n] = 0.5(0.9j)^n + 0.5(-0.9j)^n$$

To check the answer, make the following call

n = 0:1:20; x = 0.5*(0.9*j).^n + 0.5*(-0.9*j).^n
altx = filter(1,[1,0,0.81],[1,zeros(1,20)])
figure; hold on; stem(n,x,'bo'); stem(n,altx,'b*')

which results in Fig. 2.9.

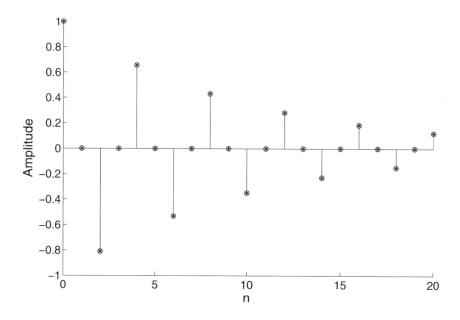

Figure 2.9: The inverse z-transform of the z-transform $1/(1 + 0.81z^{-2})$, computed using partial fraction expansion, as well as by filtering a unit impulse sequence using the b and a filter coefficients. Both results are plotted as specified by the m-code in the text.

2.5 TRANSIENT AND STEADY-STATE RESPONSES

Having discussed the Inverse-z-transform, we are now in a position to place the ideas of transient and steady-state responses on a more formal basis. To do this, we'll examine the response of a single-pole IIR to excitation by a complex exponential that begins at time zero, i.e., $n = 0$, that is to say,

$$x[n] = e^{j\omega n} u[n]$$

which has the z-transform

$$X(z) = \frac{1}{1 - e^{j\omega}z^{-1}}$$

The z-transform of a single pole LTI system represented by the difference equation

$$y[n] = b_0 x[n] + a_1 y[n-1]$$

is

$$H(z) = \frac{b_0}{1 - a_1 z^{-1}}$$

and therefore the z-transform of the system under excitation by the complex exponential is

$$Y(z) = H(z)X(z) = (\frac{b_0}{1 - a_1 z^{-1}})(\frac{1}{1 - e^{j\omega}z^{-1}})$$

which can be converted into an equivalent time domain expression using a partial fraction expansion which yields

$$y[n] = (C_S a_1^n + C_E e^{j\omega n})u[n]$$

where

$$C_S = b_0 a_1/(a_1 - e^{j\omega}); \quad C_E = b_0/(1 - a_1 e^{-j\omega})$$

Note that the excitation or driving signal, $e^{j\omega n}$ is a unity-magnitude complex exponential that does not decay away, whereas the component due to $H(z)$, namely $C_S a_1^n$, will decay away with increasing n if $|a| < 1$, i.e., if the system is stable.

Example 2.30.   Compute and plot the transient, steady state, and total responses to the complex exponential $x[n] = exp(j\pi/6)$, and independently verify the total response, of the LTI system defined by the following difference equation:

$$y[n] = x[n] - 0.85y[n-1]$$

The following code generates Fig. 2.10:

```
w = pi/6; b0 = 1; a1 = - 0.85; Cs = b0*a1/(a1-exp(j*w));
Ce = b0/(1-a1*exp(-j*w)); n = 0:1:50; figure(11);
ys = Cs*a1.^n; yE = Ce*exp(j*w*n); yZ = ys+yE;
```

subplot(321); stem(real(ys)); subplot(322); stem(imag(ys));
subplot(323); stem(real(yE)); subplot(324); stem(imag(yE));
subplot(325); stem(real(yZ)); subplot(326); stem(imag(yZ));

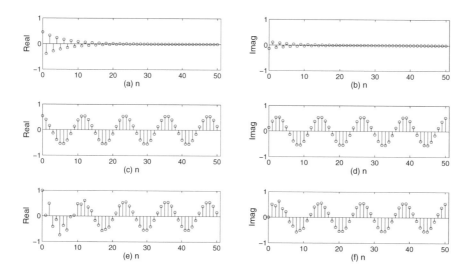

Figure 2.10: (a) Real part of transient response of IIR having a pole at -0.85 to a complex exponential at frequency pi/6; (b) Imaginary part of same; (c) Real part of steady-state response; (d) Imaginary part of steady-state response; (e) Real part of total response; (f) Imaginary part of total response.

In Fig. 2.10, we see, in plots (a) and (b), the real and imaginary parts of the transient response. Since the system pole that generates this response lies inside the unit circle, the system is stable, and this system-generated response decays away. In plots (c) and (d), we see the steady-state response due to the excitation function, a complex exponential. The total response is shown in plots (e) and (f).

We can verify the total response by filtering (say) 50 samples of the impulse response corresponding to the single pole filter having its pole at -0.85, and 50 samples of a complex exponential at radian frequency $\pi/6$. The following code results in Fig. 2.11:

```
n = [0:1:49]; y = exp(j*(pi/6)).^n;
ans = filter(1,[1,0.85],y);
figure(8); subplot(211); stem(n,real(ans));
subplot(212); stem(n,imag(ans))
```

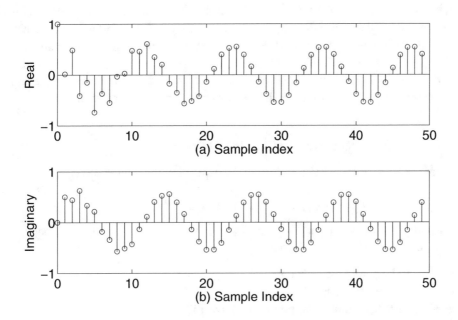

Figure 2.11: (a) Real part of complete response of an IIR having a single pole at -0.85, used to filter 50 samples of a complex exponential having radian frequency $\pi/6$; (b) Imaginary part of same.

2.6   FREQUENCY RESPONSE FROM Z-TRANSFORM

2.6.1   FOR GENERALIZED TRANSFER FUNCTION

If the Region of Convergence (ROC) of the z-transform of a digital filter includes the unit circle, the filter's frequency response can be determined by evaluating the z-transform at values of z lying on the unit circle. In other words, for each appearance of z in the z-transform, substitute a chosen value of z from the unit circle:

$$Y(e^{jw})/X(e^{jw}) = \frac{b_0 + b_1(e^{jw})^{-1}... + b_{N-1}(e^{jw})^{-(N-1)}}{1 - a_1(e^{jw})^{-1}... - a_{N-1}(e^{jw})^{-(N-1)}}$$

which simplifies to

$$Y(e^{jw})/X(e^{jw}) = \frac{b_0 + b_1 e^{-j\omega}... + b_{N-1}e^{-(N-1)j\omega}}{1 - a_1 e^{-j\omega}... - a_{N-1}e^{-(N-1)j\omega}} \qquad (2.18)$$

where ω may take on values between $-\pi$ and $+\pi$ or 0 to 2π. Equation (2.18) may then be used to obtain samples of the frequency response by substituting values for ω, computing the resulting magnitude and phase corresponding to each value of ω, and then plotting the resultant values.

Example 2.31.   A causal LTI system is defined by the z-transform below. Write an expression for the frequency response and a script to evaluate the frequency response between $-\pi$ and π.

$$H(z) = (1 + z^{-1})/(1 - 0.95z^{-1})$$

We note that the ROC is $|z| > 0.95$ and write the expression for frequency response as

$$H(e^{jw}) = (1 + e^{-jw})/(1 - 0.95e^{-jw})$$

and a script to evaluate it at values of z lying on the unit circle is as follows, with the results from the script shown in Fig. 2.12.

```
zarg = -pi:2*pi/500:pi; eaz = exp(-j*zarg);
FrqR = (1 + eaz)./(1 - 0.95*eaz);
plot(zarg/pi,abs(FrqR)); xlabel('Frequency, Units of pi')
```

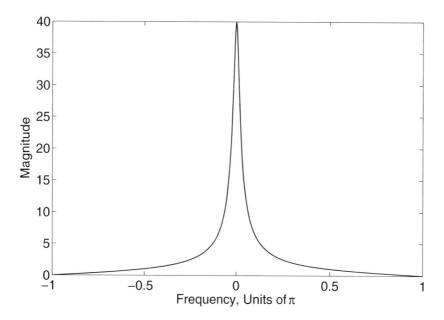

Figure 2.12: The magnitude of $H(z) = (1 + z^{-1})/(1 - 0.95\, z^{-1})$ evaluated at about 500 points along the unit circle from $-\pi$ to π radians.

2.6.2   RELATION TO DTFT

The z-transform of a sequence $x[n]$ is defined as

$$X(z) = \sum_{n=-\infty}^{\infty} x[n]z^{-n}$$

When the ROC includes the unit circle in the complex plane, we can substitute $e^{j\omega}$ for z and get

$$X(e^{j\omega}) = \sum_{n=-\infty}^{\infty} x[n]e^{-j\omega n}$$

which is the definition of the DTFT. Thus, when the unit circle is in the ROC of the z-transform of the LTI system, values of the z-transform evaluated along the unit circle are identical to values of the DTFT.

2.6.3   FINITE IMPULSE RESPONSE (FIR)

The z-transform of a finite sequence $x[n]$ is

$$x[0]z^0 + x[1]z^{-1} + x[2]z^{-2} + \dots x[N-1]z^{-(N-1)}$$

For a finite causal sequence, the **ROC** (Region Of Convergence) is everywhere in the z-plane except the origin ($z = 0$). Choosing z somewhere on the unit circle, we have

$$z = e^{j\omega}$$

We can test the response of, say, an 11-sample FIR with one-cycle correlators (equivalent to those used for Bin 1 of an 11-point DFT, for example) generated by the z-power sequence shown in plot (a) of Fig. 2.13. To test the response of the same 11-sample FIR with, say, 1.1 cycle correlators, we would use $z = \exp(j2\pi k/N)$, where $k = 1.1$.

$$b(z) = b_0 z^0 + b_1 z^{-1} + b_2 z^{-2} + \dots b_{10} z^{-10}$$

- The z-transform evaluated at a value of z lying on the unit circle at a frequency corresponding to a DTFT frequency produces the same numerical result as the DTFT for that particular frequency. We can also evaluate the z-transform at many more (a theoretically unlimited number) values of z on the unit circle, corresponding to any sample of the DTFT.

Values of z not on the unit circle may also be used, provided, of course, that they lie in the **ROC**. Values of z having magnitude less than 1.0, for example, will result in a negative power sequence of z (z^0, z^{-1}, z^{-2}, etc.) that increases in magnitude, as shown in Fig. 2.13, plots (d)-(f).

Example 2.32.   Evaluate the magnitude and phase response of the z-transform of an eight-point rectangular impulse response.

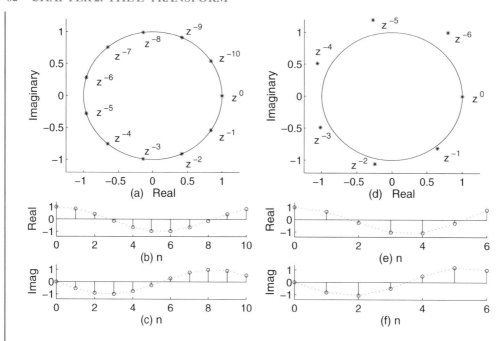

Figure 2.13: (a) Plot of complex correlator $C_1 = \exp(j2\pi/11)^n$ where $n = 0{:}{-}1{:}{-}10$; (b) Real part of C_1; (c) Imaginary part of C_1; (d) Plot of complex correlator $C_2 = (0.96\exp(j2\pi/7))^n$ where n = 0:-1:-6; (e) Real part of C_2; (f) Imaginary part of C_2.

The desired impulse response is

$$\text{Imp} = [1,1,1,1,1,1,1,1] \tag{2.19}$$

and the corresponding z-transform is

$$b(z) = 1 + z^{-1} + z^{-2} + \dots + z^{-7}$$

Let's write two scripts to do the evaluation. The first method will use a loop and evaluate the z-transform for one value of z at a time; the second uses vectorized methods that avoid looping.

For the first method, we'll create a vector of values of z along the unit circle at which to evaluate the z-transform, by use of a loop:

SR=200; for A = 1:1:SR+1; z = exp(j*(A-1)*2*pi/SR);

B = 0 :-1: -length(Imp)+1; zvec = [z.^B];

and then the z-transform for index A (i.e., at radian frequency $\pi(A-1)/SR$) is obtained by multiplying $zvec$ with the impulse response (i.e., obtaining the correlation between the impulse response and the complex correlator generated by the negative power sequence of z)

$$\text{zXform(A) = sum(Imp.*zvec); end}$$

The above uses values of z all around the unit circle, i.e., positive and negative frequencies. To plot the above with a simple 2-D plot, make the call

$$\text{plot(abs(zXform))}$$

To evaluate the frequency response at only positive frequencies, use

$$\text{z = exp(j*(A-1)*pi/SR);}$$

For the second method, we create a matrix, each column of which is the z test or evaluation vector raised to a power such as 0, -1, etc. The matrix $pzMat$ of arguments for the complex exponential is first created and then used as an argument for the exp function. Each row of the matrix $ZZMat$ is a power sequence based on a given value of z; collectively, all the rows represent power sequences for all the values of z at which the z-transform is to be evaluated.

> Imp = [1 1 1]; B=0:1:length(Imp)-1; SR = 256;
> pzMat = (2*pi*(0:1:SR)/SR)'*(-B);
> ZZMat = exp(j*pzMat); zXform = ZZMat*Imp'; plot(abs(zXform))

The script *LVxFreqRespViaZxform* (see exercises below) was used to generate Fig. 2.14 by making the call

$$\text{LVxFreqRespViaZxform([1,1,1,1,1,1,1,1],512)}$$

where the first argument is an impulse response, and the second argument is the number of z-transform and DTFT samples to compute. The figure shows, in plots (a) and (b), the magnitude and phase responses, respectively, of the z-transform of the impulse response. In the script, all values of z used to evaluate the z-transform are on the unit circle, and hence the result is identical to samples of the DTFT when evaluated at the same frequencies. The script computes the equivalent DTFT results, which are shown in plots (c) and (d) of the figure (which, for convenience shows only the positive frequency response in all plots).

Example 2.33.   Efficiently evaluate the z-transform of the FIR whose z-transform is $Y(z) = 1 + 0.9z^{-n}$ where n is large and not known ahead of time for any particular computation. If, for example, n were, say, 1000, there would be 1000 terms to evaluate, as

$$Y(z) = 1 + b_1 z^{-1} + b_2 z^{-2} + \dots + b_{1000} z^{-1000}$$

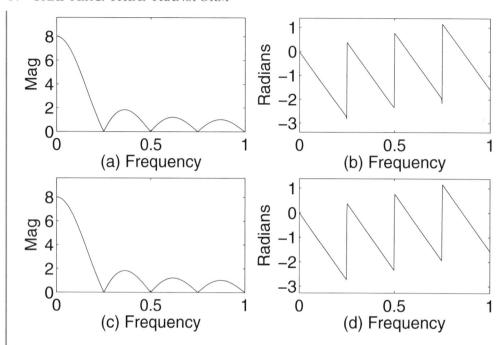

Figure 2.14: (a) Magnitude of z-Transform; (b) Phase Response of z-Transform; (c) Magnitude of DTFT; (d) Phase Response of DTFT. Note that only positive frequencies have been evaluated, i.e., only values of z lying along the upper half of the unit circle were used.

where b coefficients 1 through 999 are equal to zero.

A useful way to proceed is to assume that the vector of b coefficients will be supplied in complete form, with only a few nonzero coefficients, such as (for example)

$$b = [1, zeros(1,600), 0.7, zeros(1,600), 0.9]$$

Only nontrivial computations need to be made if we first detect the indices of the nonzero coefficients and compute only for those. The following code will work:

```
CoeffVec = [1,zeros(1,50),0.7,zeros(1,50),0.9];
SR = 256; zVec = exp(j*pi*(0:1/SR:1));
nzCoef = find(abs(CoeffVec)>0); num = zeros(1,length(zVec));
for Ctr = 1:1:length(nzCoef)
AnsThisCoeff = CoeffVec(nzCoef(Ctr))*(zVec.^ (-nzCoef(Ctr)+1));
num = num + AnsThisCoeff; end
figure(888); plot(abs(num))
```

Example 2.34.   Modify the code from the previous example to compute the z-transform of an IIR.

This can be accomplished simply by the computation **zXform = 1./num**. Let's take the simple IIR having z-transform

$$Y(z) = \frac{1}{1 - 1.27z^{-1} + 0.81z^{-2}}$$

having ROC: $|z| > 0.9$. We'll be using values of z lying on the unit circle, which is in the ROC. We thus use the following code (where we rename *num* as *den*):

```
CoeffVec = [1, -1.27, 0.81];
SR = 256; zVec = exp(j*pi*(0:1/SR:1));
nzCoef = find(abs(CoeffVec)>0); den = zeros(1,length(zVec));
for Ctr = 1:1:length(nzCoef)
AnsThisCoeff = CoeffVec(nzCoef(Ctr))*(zVec.^ (-nzCoef(Ctr)+1));
den = den + AnsThisCoeff; end
den = 1./den;
figure(777); plot(abs(den))
```

A script (see exercises below) which evaluates the magnitude of a generalized z-transform using the efficient code of the examples above for sparse coefficient vectors and which generates a 3-D plot is

$$LVxPlotZXformMagCoeff(NumCoeffVec,$$
$$...DenCoeffVec, rLim, Optr0, NSamps)$$

This script, which is intended for use with z-transforms having their ROCs lying outside a circle of radius equal to the magnitude of the largest pole of the z-transform, plots the magnitude of a z-transform that is supplied as a Numerator Coefficient Vector (*NumCoeffVec*), a Denominator Coefficient Vector (*DenomCoeffVec*), the desired number of circular contours to use (*rLim*), an optional initial contour radius, *Optr0* (pass as [] if not used), and *NSamps*, the number of frequency samples to use. When evaluating an FIR, pass *DenCoeffVec* as [1].

We can generate a 3-D plot of the magnitude of the z-transform of an eight-sample rectangular impulse response, as shown in Fig. 2.15 by making the call

LVxPlotZXformMagCoeff([1,1,1,1,1,1,1,1],[1],1,1,512)

Positive frequencies are represented by angles between 0 (normalized frequency 0) and 180 degrees (normalized frequency 1.0, or radian frequency π), measured counter-clockwise relative to the positive real axis. Negative frequencies are represented by angles between 0 and -180 degrees, or equivalently, between 180 and 360 degrees relative to the positive real axis.

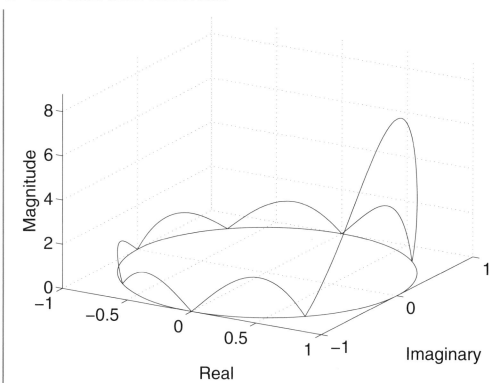

Figure 2.15: A 3-D plot of the magnitude of the z-transform of an 8-sample rectangular impulse response evaluated at many values of z along the unit circle, which is plotted in the complex plane for orientation.

Example 2.35. Evaluate the z-transform of an eight-point rectangular impulse response along contours within the unit circle.

The call

LVxPlotZXformMagCoeff([1,1,1,1,1,1,1,1],[1],30,0.95,512)

yields a figure which shows a surface formed by taking the z-transform magnitude along many circular contours in the z-plane, starting at radius 0.95 and moving outward to about radius 1.5, as shown in Fig. 2.16. A finite impulse response, which is often said to contribute only zeros to a transfer function, nonetheless has $L - 1$ trivial poles at the origin of the complex plane (L is the impulse response length), which drive the value of the z-transform to infinity at the origin, as can readily be inferred by inspection of Fig. 2.16. The plot in Fig. 2.16 was generated by evaluating the

z-transform at a limited number of points along a limited number of circular contours; as a result, fine structure, if any, between the evaluation points will be lost.

A VI that allows you to drag a zero, complex conjugate pair of zeros, or a quad of zeros (a complex conjugate pair and their reciprocals) in the z-plane, and see the magnitude and phase of the z-transform evaluated along the unit circle is

$$DemoDragZerosZxformVI$$

an example of which is shown in Fig. 2.17. This VI, when in quad mode, devolves to the minimum number of zeros necessary to maintain a linear phase filter. For example, if the main zero is given magnitude 1.0, and is complex, it and its complex conjugate are used for the FIR filter (in other words, the duplicate pair of zeros are discarded). If the main zero is given magnitude 1.0 and an angle of 0 or 180 degrees, the FIR's transfer function is built with just one zero, not four zeros. Similarly, when in Complex Conjugate mode, the number of zeros used in the FIR devolves to only one when the imaginary part of the main zero is equal to zero.

This resultant transfer function magnitude and phase responses, and the resultant impulse responses, can be compared to those obtained (in the quad case) using all four zeros in all situations if the user can run the script given below for use with MATLAB. Otherwise, the user can copy *DemoDragZerosZxformVI* to a new file name and then modify the m-code in the new VI's MathScript node to allow all four zeros to be used (see exercises below).

A script for use with MATLAB that performs functions similar to that of the VI above is

$$ML\_DragZeros$$

This script, when called, opens a GUI that allows you to select a single zero, a complex conjugate pair of zeros, or a quad of zeros (a complex conjugate pair and their reciprocals). The magnitude and phase of the z-transform as well as the real and imaginary parts of the equivalent impulse response are dynamically plotted as you move the cursor in the z-plane. This script, unlike the VI above, does not devolve to use of one or two zeros only in certain cases; four zeros are always used to create the FIR transfer function. Note that the quad of zeros always results in a linear phase characteristic and a real impulse response that is symmetrical about its midpoint. Phase linearity is of great value and FIRs can easily be designed having linear phase by ensuring that zeros of the transfer function are either single real zeros having magnitude 1.0, real pairs having reciprocal magnitudes, complex conjugate pairs having magnitude 1.0, or quads.

2.6.4   INFINITE IMPULSE RESPONSE (IIR) SINGLE POLE
The single pole IIR's z-transform

$$Y(z)/X(z) = \frac{1}{1 - pz^{-1}}$$

with ROC: $|z| > p$ (and $|p| < 1$) becomes

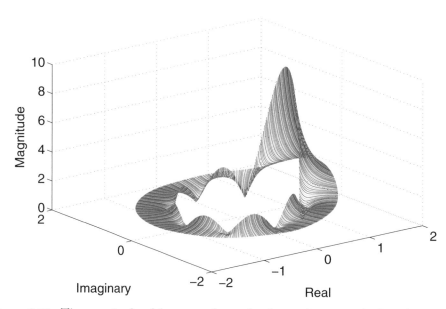

Figure 2.16: The magnitude of the z-transform of an 8-sample rectangular impulse response, evaluated over a number (30) of circular contours beginning at $|z| = 0.95$.

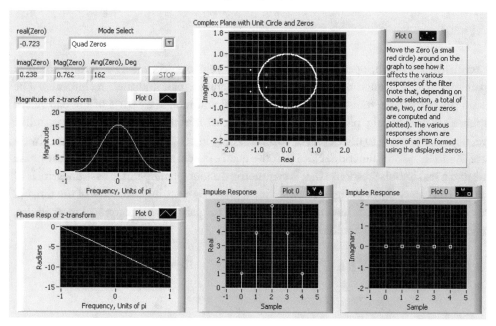

Figure 2.17: A VI that displays the magnitude and phase of the z-transform evaluated along the unit circle of an FIR filter formed from a single zero, pair of complex conjugate zeros, or a quad of zeros; the corresponding FIR impulse response is also displayed.

$$Y(z)/X(z) = \frac{1}{1 - pe^{-j\omega}}$$

after letting $z = e^{j\omega}$.

Example 2.36.   Write several lines of m-code to evaluate and plot the magnitude of the z-transform of the difference equation

$$y[n] = x[n] + 0.9y[n-1]$$

at 2000 evenly-spaced frequency samples between 0 and 2π radians.

Initially, a vector of values of z at which to evaluate the z-transform must be formed, and then a vectorized expression can be written to evaluate the z-transform at all the chosen values of z. We choose (for this example), to evaluate the z-transform at 2000 values evenly spaced along the unit circle, and thus write

```
N = 2000; zarg = 0:2*pi/(N-1):2*pi;
zVec = exp(j*zarg); DenomVec = 1 - 0.9*( zVec.^(-1));
NetzXform = 1./DenomVec;
figure(10001); plot(zarg/pi,abs(NetzXform))
xlabel('Frequency, Units of pi')
```

Example 2.37.   Determine the frequency response of the system defined by the difference equation shown below by use of the z-transform. Assume that $y[n] = x[n] = 0$ for $n < 0$. Note that the frequency response is the response to constant, unity-amplitude complex exponentials.

$$y[n] = x[n] + 1.1y[n-1]$$

This system is a single-pole IIR with a growing impulse response. The z-transform is only convergent for values of z having a magnitude greater than 1.1. Since the frequency response is determined by evaluating the z-transform for values of z on the unit circle (i.e., magnitudes of 1.0), we cannot evaluate the frequency response since the unit circle is not in the Region of Convergence.

2.6.5   CASCADED SINGLE-POLE FILTERS

If we were to cascade two identical single-pole IIRs, their equivalent z-transform would be the product of the two z-transforms, or

$$Y(z)/X(z) = (\frac{1}{1 - pz^{-1}})^2$$

Supposing we cascaded two IIRs with poles related as complex conjugates. We would have

$$Y(z)/X(z) = (\frac{1}{1 - pz^{-1}})(\frac{1}{1 - p_{cc}z^{-1}})$$

where

$$p_{cc}$$

denotes the complex conjugate of p. Doing the algebra, we get

$$Y(z)/X(z) = \frac{1}{1 - (p_{cc} + p)z^{-1} + (p_{cc}p)z^{-2}}$$

which reduces to

$$Y(z)/X(z) = \frac{1}{1 - (2 \cdot \text{real}(p))z^{-1} + |p|^2 z^{-2}} \tag{2.20}$$

Example 2.38.   Evaluate the magnitude of the z-transform for a cascaded connection of two single-pole IIR filters each having a pole at 0.9.

Using Eq. (2.20), the z-transform would be:

$$H(z) = \frac{1}{1 - 1.8z^{-1} + 0.81z^{-2}} \tag{2.21}$$

The z-transform of Eq. (2.21), evaluated along a constant radius contour having radius 1.0 (i.e., the unit circle) is shown in Fig. 2.18. It was obtained by making the call

LVxPlotZXformMagCoeff([1],[1,-1.8,0.81],[1],[1],256)

Example 2.39.   Evaluate the z-transform of a system consisting of two poles at $0.9j$ and $-0.9j$.

These poles resonate at the half-band frequency, and, using Eq. (2.20), the net z-transform is

$$\frac{1}{1 + 0.81z^{-2}}$$

Figure 2.19, which shows a peaked frequency response at the positive and negative half-band frequencies, was obtained by making the call

LVxPlotZXformMagCoeff([1],[1,0,0.81],[1],[1],256)

Example 2.40.   Evaluate the magnitude of the z-transform along the unit circle of a system consisting of a pole at $(0.5 + 0.85j)$ and its complex conjugate.

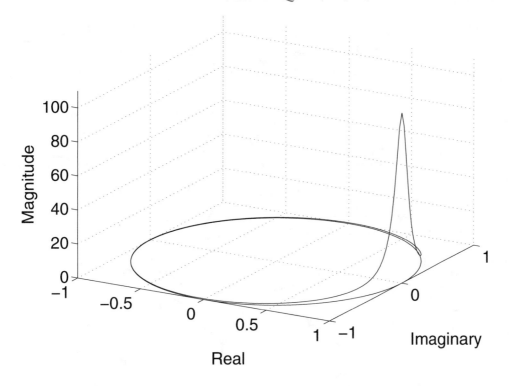

Figure 2.18: The magnitude of the z-transform evaluated along the unit circle (i.e., the Frequency Response) for the cascaded connection of two simple IIRs each having a real pole at 0.9 (the unit circle is shown in the complex plane for reference).

Again using Eq. (2.20),we get the z-transform as

$$Y(z)/X(z) = \frac{1}{1 - (2 \cdot 0.5)z^{-1} + (0.5^2 + 0.85^2)z^{-2}}$$

The z-transform is then evaluated as discussed in previous examples by substituting values of z on the unit circle.

A script that is equivalent to *LVxPlotZXformMagCoeff* and which works well with MATLAB (and which plots the unit circle underneath the z-transform magnitude as shown, for example, in Fig. 2.19) is

$$MLPlotZXformMagCoeff(NCoeffVec, DCoeffVec, rLim, Optr0, NSamps)$$

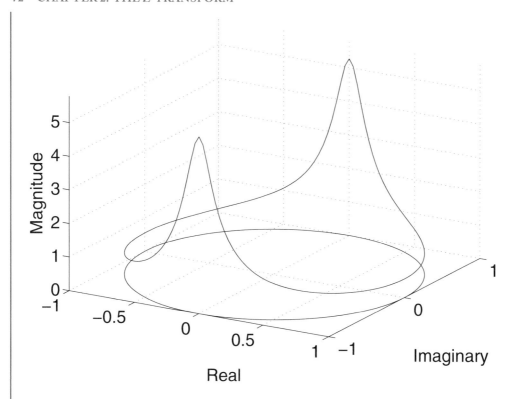

Figure 2.19: The magnitude of the *z*-transform evaluated along the unit circle (i.e., the Frequency Response) for two cascaded single-pole IIRs, the two poles being complex conjugates, tuned to the half-band frequency.

where $NCoeffVec$ is the numerator coefficient vector (i.e., *b* coefficients) and $DCoeffVec$ is the denominator coefficient vector (i.e., the *a* coefficients). For a description of all arguments and sample calls, type the following on the MATLAB Command Line and then press Return:

help MLPlotZXformMagCoeff

Another script

$$LVxPlotZTransformMag(ZeroVec, PoleVec, rMin, rMax, NSamps)$$

will take arguments as a row vector of zeros (*ZeroVec*), a row vector of poles (*PoleVec*), the minimum radius contour to compute (*rMin*), the maximum radius contour to compute (*rMax*), and the number of frequency samples along each contour to compute, *NSamps*. By passing both *rMin* and *rMax* as 1, the standard frequency response can be obtained.

A call to solve the current example would be

LVxPlotZTransformMag([],[(0.5 + 0.85*j),(0.5 - 0.85*j)],1,1,256)

which results in Fig. 2.20.

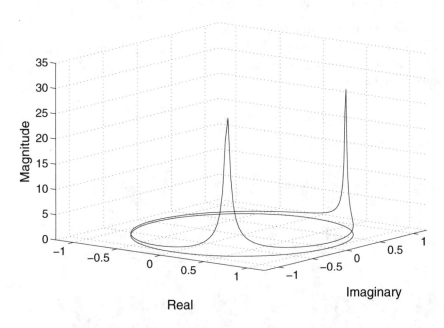

Figure 2.20: Magnitude of z-transform of the IIR formed from the poles (0.5 + 0.85j) and (0.5 - 0.85j), evaluated along the unit circle.

Example 2.41. Evaluate the frequency response of the z-transform below at radian frequency $\pi/4$, and then obtain the impulse response for the first 150 samples, correlate it with the complex exponential corresponding to the same radian frequency (i.e., compute the CZL), and compare the correlation value to the answer obtained directly from the z-transform. The z-transform is

$$X(z) = \frac{1}{1 - 0.9z^{-1}}$$

For the direct evaluation, we substitute exp(j*pi/4) for z in the z-transform and make the following call:

ansZD = 1./(1 - 0.9*exp(-j*pi/4))

which yields $ansZD$ = 0.6768 -1.1846i. To estimate the frequency response using time domain correlation, we obtain the impulse response (for the first 150 points, which should be adequate since it decays quickly), and correlate with the first 150 samples of the complex exponential power series of **exp(-j*pi/4)**

$$ansTD = sum((0.9.\hat{}(0:1:149)).*(exp(-j*pi/4). \hat{}(0:1:149)))$$

which produces the identical answer to six decimal places.

A VI that allows you to drag a pole or pair of complex conjugate poles in the z-plane while observing the magnitude and phase responses of the z-transform, as well as the real and imaginary parts of the equivalent impulse response, is

$$DemoDragPolesZxformVI$$

an example of which is shown in Fig. 2.21.

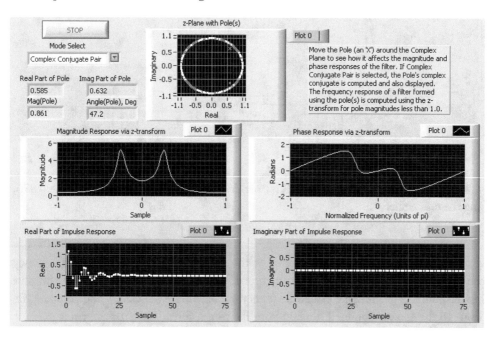

Figure 2.21: A VI that allows the user to drag a pole or pair of complex conjugate poles in the z-plane and observe the effect on impulse response and z-transform of an IIR formed from the pole(s).

2.6.6   OFF-UNIT-CIRCLE ZEROS AND DECAYING SIGNALS

When transfer function zeros are off the unit circle in the z-plane, their effect on the transfer function (for steady-state complex exponentials) is diminished. Such "off-unit-circle zeros" do not drive the output of the system to zero for a steady-state, unity-magnitude complex exponential, but rather for one which has a decaying (for zeros inside the unit circle) or growing (for zeros outside the unit circle) magnitude with time.

Example 2.42.   Determine, for an FIR having the following transfer function

$$1 - 0.9z^{-1} \tag{2.22}$$

what input signal will result in a zero-valued output signal.

We can set expression (2.22) to zero, i.e.,

$$1 - 0.9z^{-1} = 0 \tag{2.23}$$

and solving for z we get $z = 0.9$.

To see this in the time domain, consider the decaying exponential series $x[n] = 1, 0.9, 0.81,$ ..., i.e.,

$$x[n] = 0.9^n$$

where $0 \leq n \leq \infty$. Let's compute the first few terms of the response of the single zero system whose z-transform is shown at (2.22) and whose difference equation is

$$y[n] = x[n] - 0.9x[n-1] \tag{2.24}$$

Since Eq. (2.24) represents an FIR, the impulse response is identical to the FIR coefficients, which are [1, -0.9]. We can then convolve $x[n]$ with the system impulse response to obtain the system output $y[n]$:

n = 0:1:20; x = 0.9.^n;
y = conv([1, -0.9],x)
stem(y)

Note that the first and last samples of the above computation are not zero–why?

Example 2.43.   Compute the output for an FIR with one zero at 0.9 and a constant amplitude signal having the same frequency as that of the zero, i.e., DC.

We can use the same call as above, but with the signal amplitude constant:

n = 0:1:20; x = 1.^n;
y = conv([1, -0.9],x)
stem(y)

or, alternatively,

$$\text{stem(filter([1,-0.9],[1],[1.^(0:1:50)]))}$$

The output sequence is 1 followed by the steady-state value of 0.1 and the final transient (non-steady-state) value as the end of $x[n]$ is reached.

2.7   TRANSFER FUNCTION & FILTER TOPOLOGY

2.7.1   DIRECT FORM

Consider the difference equation

$$y[n] = b_0 x[n] + b_1 x[n-1] + b_2 x[n-2] - a_1 y[n-1] - a_2 y[n-2]$$

or

$$y[n] + a_1 y[n-1] + a_2 y[n-2] = b_0 x[n] + b_1 x[n-1] + b_2 x[n-2]$$

and its z-transform:

$$Y(z)/X(z) = \frac{b_0 + b_1 z^{-1} + b_2 z^{-2}}{1 + a_1 z^{-1} + a_2 z^{-2}}$$

which can be written as a product of two terms

$$Y(z)/X(z) = (\frac{1}{1 + a_1 z^{-1} + a_2 z^{-2}})(\frac{b_0 + b_1 z^{-1} + b_2 z^{-2}}{1})$$

each of which can be realized as a separate structure, and the two structures cascaded, as shown in Fig. 2.22. This arrangement is known as a Direct Form I realization. Note that the numerator and denominator of the z-transform are separately realized and the two resultant time domain filters are cascaded, which provides the time domain equivalent (convolution) of the product of the two transfer functions.

Note that in Fig. 2.22, the two delay chains come from the same node, and hence may be replaced by a single delay chain, resulting in a Direct Form II realization as shown in Fig. 2.23. As many delay stages as needed may be used in a Direct form filter realization, as shown in Fig. 2.24.

A somewhat simpler method of illustrating a filter's topology is a signal flow diagram, an example of which is shown in Fig. 2.25, which, like Fig. 2.24, depicts an m-th order Direct Form II filter.

2.7.2   DIRECT FORM TRANSPOSED

The exact same transfer function as depicted in Fig. 2.26, for example, can be realized by following the simple procedure of: 1) reversing the directions of all signal flow arrows; 2) exchanging input and output. The result is a Direct Form Transposed structure.

2.7.3   CASCADE FORM

In addition to the Direct Form, there are several alternative structures for realizing a given transfer function. When the order is above two, it is often convenient to realize the net transfer function as a cascade of first and second order sections, each second order section being known as a **Biquad**. The z-transform is then a product of second order z-transforms with possibly an additional first order z-transform. The general form is illustrated in Fig. 2.27.

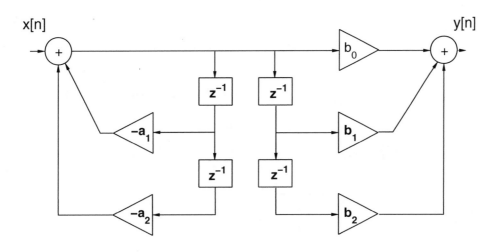

Figure 2.22: A Direct Form I realization of a second-order LTI system.

Thus, the N-th order z-transform

$$H(z) = \frac{b_0 + b_1 z^{-1} + ...b_N z^{-N}}{1 + a_1 z^{-1} + ...a_N z^{-N}}$$

may be converted into the equivalent form

$$H(z) = b_0 \prod_{k=1}^{K} \frac{1 + B_{k,1} z^{-1} + B_{k,2} z^{-2}}{1 + A_{k,1} z^{-1} + A_{k,2} z^{-2}}$$

where $K = N/2$ when N is even, or $(N-1)/2$ when N is odd, and the A and B coefficients are real numbers. When the order is even, there are only second order sections, and when it is odd there is one first order section in addition to any second order sections. Most filter designs generate poles in complex conjugate pairs, with perhaps one or more single, real poles, and therefore making each second order section from complex conjugate pairs and/or real pairs results in second order sections having all-real coefficients.

Referring to the function *LVDirToCascade*, the input arguments b and a are the numerator and denominator coefficients of a Direct Form filter; while the output arguments *Ac* and *Bc* are the biquad section coefficients, and *Gain* is the corresponding gain, the computation of which is shown in the m-code following the next paragraph.

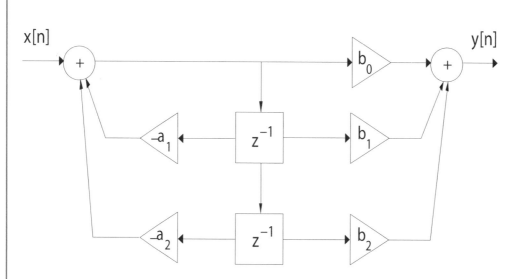

Figure 2.23: A Direct Form II realization of a second-order LTI system.

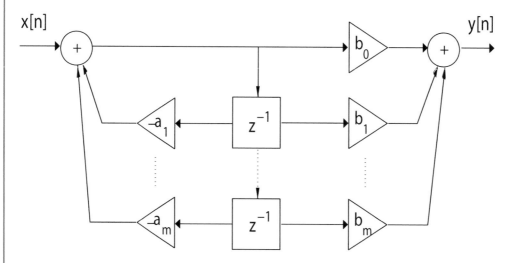

Figure 2.24: A Direct Form II realization of an m-th order LTI system.

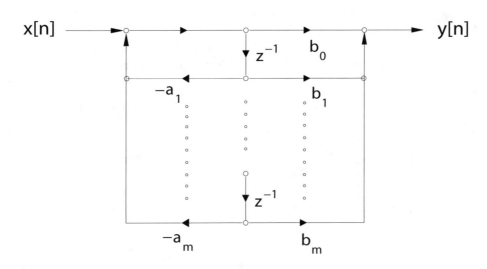

Figure 2.25: Signal Flow diagram for a generalized m-th order Direct Form II filter structure or realization. Signal flow direction is depicted by small arrows, which have unity gain unless otherwise labeled. Typical labels include a number representing a gain, a variable representing a gain, or the symbol z^{-1}, which represents a delay of one sample. The summing of signals is performed when any two signals flowing in the same direction meet at a node, represented by a small circle. The reader should compare this figure to the previous one.

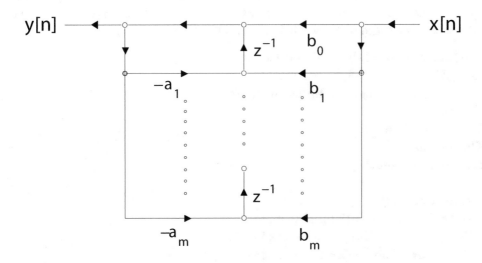

Figure 2.26: The signal flow diagram for an m-th order Direct Form II Transposed filter structure.

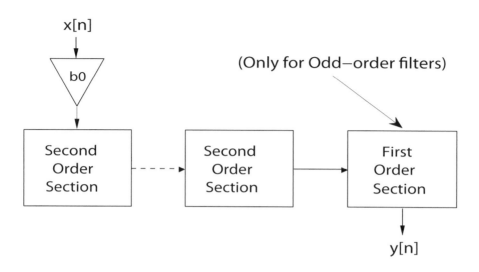

Figure 2.27: A Cascade Form filter arrangement, in which pairs of complex conjugate poles and pairs of real poles have been collected to form second order sections having real coefficients; if the order is odd, a single real pole will be left over to form a first order section.

The heart of a script to convert from Direct to Cascade form is a function to place the poles and/or zeros in ordered pairs, thus making it easy to compute biquad section coefficients. The function *LVCmplxConjOrd*, which is set forth below, fulfills this function.

```
function [Bc,Ac,Gain] = LVDirToCascade(b,a)
B0 = b(1); A0 = a(1); a = a/A0; b = b/B0;
Gain = B0/A0; lenB = length(b); lenA = length(a);
if lenB<lenA
b = [b,zeros(1,lenA-lenB)];
elseif lenA<lenB
a = [a,zeros(1,lenB-lenA)];
end
Ord = length(roots(b));
if Ord-2*fix(Ord/2) > 0; b = [b,0]; a = [a,0]; end
[NumRtsB] = LVCmplxConjOrd(roots(b),10^(-10));
[NumRtsA] = LVCmplxConjOrd(roots(a),10^(-10));
LenNumRts = length(NumRtsB); NoPrs = fix(LenNumRts/2)
for Ctr = 1:1:NoPrs; ind = [2*Ctr-1, 2*Ctr];
```

```
Bc(Ctr,1:3) = real(poly(NumRtsB(ind)));
Ac(Ctr,1:3) = real(poly(NumRtsA(ind)));
end
```

The function *LVCmplxConjOrd*, used in the script *LVDirToCascade* above, sorts a list of randomly-ordered complex conjugate poles and real poles into pairs of complex conjugates followed by real-only poles: x is a vector of complex conjugate poles and real poles, in random (or unknown) order, *tol* defines how close a pole must be to its conjugate to be detected as such (*tol* is necessary due to roundoff error, which may prevent true conjugates from being so detected) *CCPrs* is the output vector of pairs of complex conjugates, with real poles at the trailing end.

```
function [CCPrs] = LVCmplxConjOrd(x,tol)
Lenx = length(x); CCPrs = x;
for ctr = 1:2:Lenx-1
absdiff = abs(CCPrs(ctr)-conj(CCPrs(ctr+1:Lenx)));
y = find( absdiff < tol );
if isempty(y) % real number
temp = CCPrs(ctr); CCPrs(ctr:Lenx-1) = CCPrs(ctr+1:Lenx);
CCPrs(Lenx) = temp;
else; temp = CCPrs(ctr+1); CCPrs(ctr+1) = CCPrs(y(1)+ctr);
CCPrs(y(1)+ctr) = temp; end; end
```

To perform filtering using the Cascade Form filter coefficients, the input signal is filtered using the first cascade filter section, the output of which is used as the input to the second cascade filter section, and so on, until the signal has passed through all of the cascaded filter sections. The script

$$y = LVCascadeFormFilter(Bc, Ac, Gain, x)$$

works in this manner to filter the input vector x to produce the net filtered output y:

```
function [y] = LVCascadeFormFilter(Bc,Ac,Gain,x)
szA = size(Ac);
for ctr = 1:1:szA(1)
a = Ac(ctr,:); b = Bc(ctr,:);
x = filter(b,a,x);
end
y = Gain*x;
```

Example 2.44. Write a test script that will begin with a set of Direct Form filter coefficients, convert to Cascade Form, filter a linear chirp first with the Direct Form coefficients, then (as a separate experiment), filter a linear chirp using the Cascade Form coefficients, and plot the results from both experiments to show that the results are identical.

```
[b,a] = butter(7,0.4),
x = chirp([0:1/1023:1],0,1,512);
y1 = filter(b,a,x);
[Bc,Ac,Gain] = LVDirToCascade(b,a),
[y2] = LVCascadeFormFilter(Bc,Ac,Gain,x);
figure(10); subplot(211); plot(y1);
subplot(212); plot(y2)
```

To convert from Cascade to Direct form, a script must separately convolve each of the numerator factors and each of the denominator factors. Ac is a K by 3 matrix of biquad section denominator coefficients, K being $N/2$ for even filter order N, or $(N + 1)/2$ for odd filter order, Bc is an r by 3 matrix of biquad section numerator coefficients, with $r = \text{floor}(N/2)$ where N is the filter order.

```
function [b,a,k] = LVCas2Dir(Bc,Ac,Gain)
k = Gain; szB = size(Bc);
b = 1; for ctr = 1:1:szB(1),
b = conv(b,Bc(ctr,:)); end;
szA = size(Ac); a = 1;
for ctr = 1:1:szA(1),
a = conv(a,Ac(ctr,:)); end
```

Example 2.45.   Use the call $[b, a] = butter(4, 0.5)$ to obtain a set of Direct Form Butterworth filter coefficients, convert to Cascade Form, and then convert from Cascade back to Direct Form.

Using the scripts given above, a composite script is

```
[b,a] = butter(4,0.5)
[Bc,Ac,Gain] = LVDirToCascade(b,a)
[b,a,k] = LVCas2Dir(Bc,Ac,Gain)
```

which yields, from the first line,

$$b = [0.094, 0.3759, 0.5639, 0.3759, 0.094]$$

which can be rewritten as

$$0.094([1,4,6,4,1])$$

and

$$a = [1,0,0.486,0,0.0177]$$

The second m-code line above yields (in compact reformatted form) the Cascade Form coefficients as

$$Bc = [1,2,1;1,2,1]$$

$$Ac = [1,0,0.4465;1,0,0.0396]$$

The third m-code line above converts the Cascade Form coefficients back into Direct Form:

$$b = [1,4,6,4,1]$$

$$a = [1,0,0.4860,0,0.0177]$$

$$k = 0.094$$

2.7.4 PARALLEL FORM

In this form, illustrated in Fig. 2.28, a number of filter sections are fed the input signal in parallel, and the net output signal is formed as the sum of the output signals from each of the sections. Eq. (2.13) shows a Direct Form z-transform, possibly having the numerator order exceeding that of the denominator, and Eq. (2.14) rewrites this as a proper fraction (i.e., having numerator order equal to or less than that of the denominator) plus a polynomial in z. In the Parallel Form, Eq. (2.14) is rewritten as

$$H(z) = \frac{B_{k,0} + B_{k,1}z^{-1}}{1 + A_{k,1}z^{-1} + A_{k,2}z^{-2}} + \sum_{0}^{M-N} C_k z^{-k}$$

where M and N are the orders of the numerator and denominator of the z-transform, respectively, and, assuming N is even, $K = N/2$, and the A, B, and C coefficients are real numbers. Note that in this form, the numerator of the second order sections is first order.

To generate the Parallel Form filter coefficients Bp, Ap, and Cp, the script (see exercises below)

$$[Bp, Ap, Cp] = LVxDir2Parallel(b, a)$$

receives as input the Direct Form filter coefficients (b, a) and begins by using the function *residuez* to produce the basic vectors of residues, poles, and C coefficients, as described earlier in the chapter with respect to partial fraction expansion. It is then necessary to reorder the poles in complex conjugate pairs, followed by real poles. This is done using the function *LVCmplxConjOrd*, introduced above with respect to conversion to Cascade Form. Since it is necessary to keep a given residue with its corresponding pole, it is necessary, once the properly-ordered vector of poles has been obtained, to locate each one's index in the original pole vector returned by the call to *residuez*, and then reorder the residues obtained from that call to correspond to the new pole order obtained from the call to

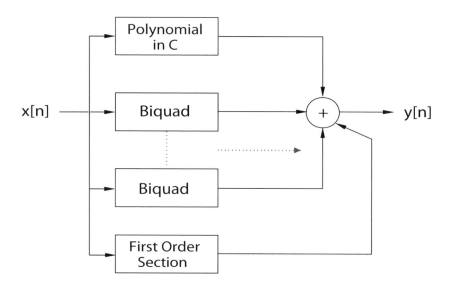

Figure 2.28: A Parallel Form Filter. Assuming even order N, there will be $K = N/2$ biquad sections. Each biquad section has a second-order denominator and a first-order numerator. If any of the C coefficients are nonzero, the section labeled "Polynomial in C" will be present. If N is odd, there will be one additional first-order section in parallel with the other sections, as shown.

LVCmplxConjOrd. Once this has been done, the poles and corresponding residues are collected two at a time and a call is made to *residuez*, which in this mode delivers a set of b and a coefficients for a single biquad section. This is done until all pole pairs have been used, and if there is one pole and residue left over, a single first order section is also created.

In order to filter using the Parallel Form, the input signal is convolved with each filter section, and the outputs of the sections are added. The script (see exercises below)

$$y = LVxFilterParallelForm(Bp, Ap, Cp, x)$$

filters the input vector x using the Parallel Form coefficients (Bp, Ap, Cp) to produce the output signal y.

Example 2.46.   Write a short script that generates a basic set of Direct Form coefficients, filters a linear chirp using them with the function filter, and then converts the Direct Form coefficients into Parallel Form coefficients, filters the linear chirp using the script *LVxFilterParallelForm*, and then displays plots of the output signals from the two filtering operations.

The following code results in Fig. 2.29.

```
[b,a] = butter(5,0.2)
x = chirp([0:1/1023:1],0,1,512);
y1 = filter(b,a,x);
figure(9); subplot(211); plot(y1);
[Bp,Ap,Cp] = LVxDir2Parallel(b,a);
y2 = LVxFilterParallelForm(Bp,Ap,Cp,x);
subplot(212); plot(y2)
```

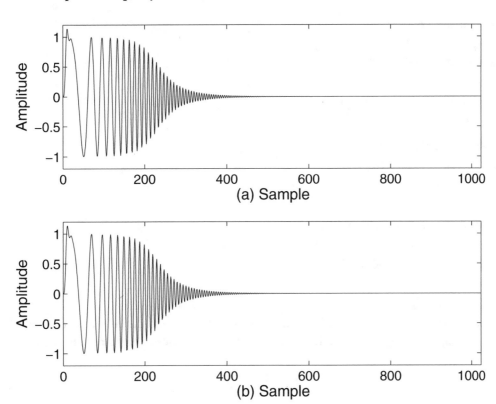

Figure 2.29: (a) A linear chirp filtered using Direct Form coefficients of a fifth order lowpass Butterworth filter; (b) The same linear chirp, filtered by the equivalent Parallel Form coefficients of the fifth order lowpass Butterworth filter.

In order to convert from Parallel Form to Direct Form, the script (see exercises below)

$$[b, a] = LVxParallel2Dir(Bp, Ap, Cp)$$

uses the coefficients for each filter section to obtain the corresponding set of poles and residues by making a call to *residuez*. This is done for all Parallel Form filter sections, and the entire accumulated set of poles and residues is used in a single "reverse" call to *residuez* to obtain the Direct Form *b* and *a* coefficients.

Example 2.47.   Use the call $[b, a] = butter(3, 0.4)$ to obtain a set of coefficients (b, a) for a lowpass filter, convert to a set of Parallel Form coefficients $[Bp, Ap, Cp]$, and then convert these coefficients back into Direct Form coefficients using the scripts described above.

A suitable "composite" script would be

[b,a] = butter(3,0.4)
[Bp,Ap,Cp] = LVxDir2Parallel(b,a)
[b,a] = LVxParallel2Dir(Bp,Ap,Cp)

The first m-code line above results in

b = [0.0985, 0.2956, 0.2956, 0.0985]
a = [1, -0.5772, 0.4218, -0.0563]

The second line results in

Bp = [-1.229, 0.3798; 3.0777, 0]
Ap = [1, -0.4189, 0.3554]
Cp = [-1.7502]

The third m-code line, as expected, yields

b = [0.0985, 0.2956, 0.2956, 0.0985]
a = [1, -0.5772, 0.4218, -0.0563]

2.7.5   LATTICE FORM

Three basic lattice forms are the All-Zero Lattice, the All-Pole Lattice, and the Pole-Zero Lattice or Lattice-Ladder. Figure 2.30 shows the basic topology of an all-zero lattice, suitable for implementing an FIR.

The All-Zero Lattice is characterized by the two equations

$$g_m[n] = g_{m-1}[n] + K_m h_{m-1}[n-1], \; m = 1, 2, ...M - 1$$

$$h_m[n] = K_m g_{m-1}[n] + h_{m-1}[n-1], \; m = 1, 2, ...M - 1$$

where the values of K are known as the reflection coefficients. For a given FIR in Direct Form, a recursive algorithm can be solved to determine the appropriate values of K. However, if the magnitude of any value of K_m is 1, the algorithm will fail. This is the case for linear phase filters, so linear phase filters cannot be implemented using a lattice structure.

Example 2.48.   For a Direct Form FIR having $b = [1, 0.5, -0.8]$, determine the equivalent K coefficients by writing the lattice transfer function and equating to b.

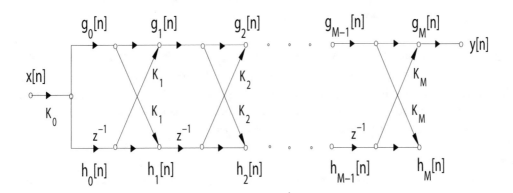

Figure 2.30: The basic structure of an All-Zero Lattice, suitable to implement an FIR.

Referring to Fig. 2.30, we would need two stages, so in this case the output would be at $g_2[n]$. By tracing the three signal paths from the input to the output at $g_2[n]$, we can write the lattice output transfer function as

$$Y(z)/X(z) = 1 + (K_1 + K_1 K_2)z^{-1} + K_2 z^{-2}$$

and since the Direct Form z-transform is

$$Y(z)/X(z) = b[0] + b[1]z^{-1} + b[2]z^{-2}$$

we get

$$(K_1 + K_1 K_2) = b[1]; \; K_2 = b[2]$$

or

$$K_2 = b[2]; \; K_1 = b[1]/(1 + b[2])$$

We thus get $K_1 = 0.5/(1\text{-}0.8) = 2.5$ and $K_2 = -0.8$. We can check this using the following call to the MathScript function *tf2latc*:

$$\textbf{k = tf2latc([1,0.5,-0.8])}$$

which yields k = 2.5000, -0.8000. Note that the above derivation assumes $b[0]$ = 1, so b should be scaled by its first element, i.e., **b = b/b[0]**.

The general recursive algorithm to compute all-zero lattice coefficients for any order of FIR may be found in references [4] and [5].

Example 2.49. Use the MathScript function $k = tf2latc(b)$ to determine the K parameters for the FIR having b = [2, 1, 0, 1.5]. Use the MathScript function $[g, h] = latcfilt(k, x)$ to filter the signal $x[n]$, which can be a linear chirp. Also filter the chirp using the Direct Form b coefficients, and plot the results from both filtering operations. Finally, make the call $b = latc2tf(k)$ to convert the lattice k coefficients back into Direct Form (b) coefficients.

We run the following m-code, in which we process a linear chirp with the Direct Form FIR b coefficients, and then using the K coefficients in an All-Zero Lattice, the results of which are shown in Fig. 2.31.

```
x = chirp([0:1/1023:1],0,1,512);
b = [2,1,0,1.5]; b1 = b(1); b = b/b1; y = filter(b,1,x);
k = tf2latc(b), [g,h] = latcfilt(k,x);
figure(8); subplot(211); plot(y); xlabel('Sample')
subplot(212); plot(g); xlabel('Sample')
b = latc2tf(k); b = b*b1
```

The topology of an All-Pole Lattice is shown in Fig. 2.32; note that the reflection coefficients are feedback rather than feedforward (see reference [4] for details). The MathScript function $tf2latc$ can be used to determine the values of K using the format $K = tf2latc(1, a)$ where the input argument a represents the Direct Form IIR coefficients. To filter a sequence x[n] using the all-pole lattice coefficients k, make the following call: $[g, h] = latcfilt(k, 1, x)$.

Example 2.50. A certain IIR has a = [1, 1.3, 0.81]. Compute the All-Pole Lattice coefficients K and filter a linear chirp using both the Direct Form and Lattice coefficients. Plot the results from each filtering operation.

The following code will suffice, the results of which are shown in Fig. 2.33.

```
x = chirp([0:1/1023:1],0,1,512);
a = [1,1.3,0.81]; y = filter(1,a,x);
k = tf2latc(1,a),
[g,h] = latcfilt(k,1,x);
figure(8); subplot(211); plot(y);
subplot(212); plot(g)
```

A filter having both IIR and FIR components has an equivalent in the Lattice-Ladder, the basic design of which is shown in Fig. 2.34. The call

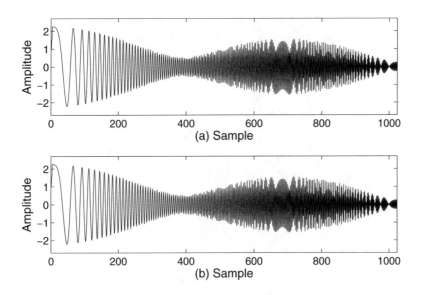

Figure 2.31: (a) A linear chirp after being filtered using the Direct Form FIR coefficients [2,1,0,1.5]; (b) A linear chirp after being filtered using the equivalent All-Zero Lattice structure.

$$[K, C] = tf2latc(b, a)$$

computes the lattice coefficients K and the ladder coefficients C that form the equivalent Lattice-Ladder to the Direct Form filter characterized by the z-transform numerator coefficients b and denominator coefficients a. To filter a signal $x[n]$ using the lattice and ladder coefficients K and C, respectively, make the call

$$[g, h] = latcfilt(K, C, x)$$

Example 2.51. A certain IIR has Direct Form coefficients b = [0.3631, 0.7263, 0.3631] and a = [1, 0.2619, 0.2767]. Compute the lattice and ladder coefficients K and C, respectively, and filter a linear chirp using both the Direct Form and Lattice-Ladder coefficients.

The results from the following code are shown in Fig. 2.35.

```
x = chirp([0:1/1023:1],0,1,512);
a = [1,0.2619,0.2767];
b = [0.3631,0.7263,0.3631];
y = filter(b,a,x);
```

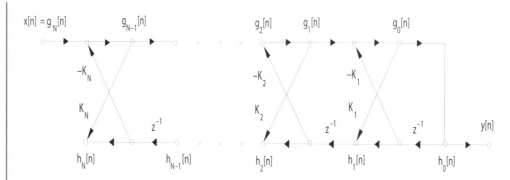

Figure 2.32: The general topology of an All-Pole Lattice filter.

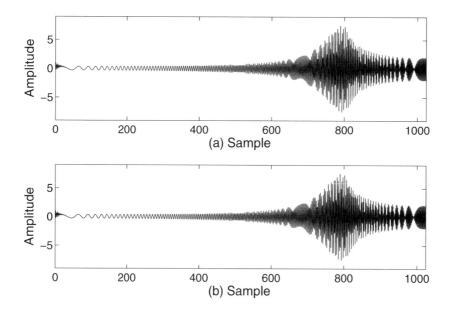

Figure 2.33: (a) A linear chirp after being filtered using the Direct Form IIR coefficients b =1 and a = [1,1.3,0.81]; (b) A linear chirp after being filtered using an All-Pole Lattice having k = [0.7182, 0.81].

```
[k,c] = tf2latc(b,a),
[g,h] = latcfilt(k,c,x);
figure(8); subplot(211); plot(y);
subplot(212); plot(g);
```

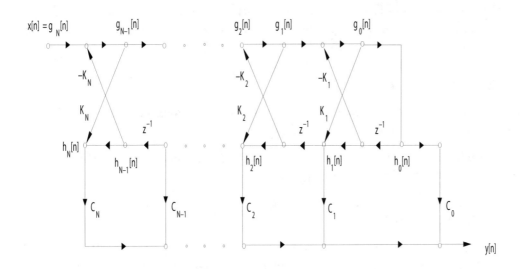

Figure 2.34: A generalized Lattice-Ladder structure.

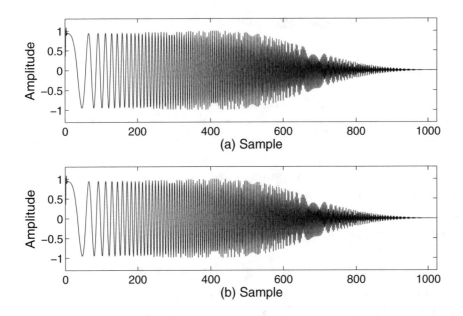

Figure 2.35: (a) A linear chirp filtered by an IIR having $b = [0.3631, 0.7263, 0.3631]$ and $a = [1, 0.2619, 0.2767]$; (b) A linear chirp filtered using a Lattice-Ladder structure having $k = [0.2051, 0.2767]$ and $c = [0.1331, 0.6312, 0.3631]$.

Realizations specific to FIRs only (Direct or Transversal, Linear Phase, Cascade, and Frequency Sampling Forms) are discussed in Volume III of the series (see the Preface to this volume for information on the contents of Volume III).

2.8   REFERENCES

[1] James H. McClellan et al, *Computer-Based Exercises for Signal Processing Using MATLAB 5*, Prentice-Hall, Upper Saddle River, New Jersey, 1998.

[2] James H. McClellan, Ronald W. Schaefer, and Mark A. Yoder, *Signal Processing First*, Pearson Prentice Hall, Upper Saddle River, New Jersey, 2003.

[3] Alan V. Oppenheim and Ronald W. Schaefer, *Discrete-Time Signal Processing*, Prentice-Hall, Englewood Cliffs, New Jersey, 1989.

[4] John G. Proakis and Dimitris G. Manolakis, *Digital Signal Processing, Principles, Algorithms, and Applications, Third Edition*, Prentice Hall, Upper Saddle River, New Jersey, 1996.

[5] Vinay K. Ingle and John G. Proakis, *Digital Signal Processing Using MATLAB V.4*, PWS Publishing Company, Boston, 1997.

[6] Richard G. Lyons, *Understanding Digital Signal Processing, Second Edition*, Prentice Hall, Upper Saddle River, New Jersey 2004.

2.9   EXERCISES

1. Determine the z-transform and the ROC for the following sequences:

      (a) $n = [0{:}1{:}5]$; $x[n] = [2,0,1,-1,0,1,8]$
      (b) $x[n] = 0.95u[n]$
      (c) $x[n] = 2\delta[n+1] + \delta[n] + 0.8u[n]$
      (d) $x[n] = -0.5\delta[n] - \delta[n-1](0.8) + 0.8u[n]$
      (e) $n = [2{:}1{:}6]$; $x[n] = [-1,0,1,5,-3]$
      (f) $x[n] = (n-2)u[n-2]$
      (g) $x[n] = 0.75^{n-1}u[n-1] + 0.95^n u[-n-1]$
      (h) $x[n] = 2.5^n u[n-1]$

2. Determine the z-transform and ROC corresponding to each the following difference equations:

      (a) $y[n] = x[n] + 0.75x[n]$
      (b) $y[n] = 0.5x[n] + x[n-1] - 2x[n-2] + x[n-3] + 0.5x[n-4]$
      (c) $y[n] = 0.75x[n] - 2x[n-1] + 3x[n-2] + 1.3y[n-1] - 0.95y[n-2]$
      (d) $y[n] = x[n] - 1.1x[n-1] - 2.3x[n-2] - 0.98y[n-2]$

3. Since two time domain sequences may be equivalently convolved by multiplying their z-transforms, correlation can be equivalently performed by multiplying the z-transform of one sequence and the

z-transform of a time-reversed version of the other sequence. The time reversal can be obtained by substituting $1/z$ (see the time reversal property of the z-transform earlier in the chapter) for z in the z-transform of the sequence to be reversed. Perform correlation of the following pairs of sequences $x_1[n]$ and $x_2[n]$ by obtaining the time domain sequence corresponding to the product of the z-transform of $x_1[n]$ and $x_2[-n]$, then verify the answer by using time domain correlation. Don't forget to take note of the applicable ROCs.

(a) $x_1[n] = [1,2,3,4]$; $x_2[n] = [4,3,2,1]$; $n = [0,1,2,3]$
(b) $x_1[n] = [1,2,3,4]$; $x_2[n] = [1,2,3,4]$; $n = [0,1,2,3]$
(c) $x_1[n] = 0.9(u[n] - u[n - 5])$; $x_2[n] = 0.8(u[n] + u[n - 3])$

4. Determine the poles and zeros corresponding to each of the following z-transforms, and, using paper and pencil, plot them in the z-plane. Characterize each system's stability.

(a) $X(z) = (1 + z^{-1}) / (1 - 1.4z^{-1} + 0.9z^{-2})$
(b) $X(z) = (1 + z^{-3}) / (1 + 0.9z^{-3})$
(c) $X(z) = (0.0528 + 0.2639z^{-1} + 0.5279z^{-2} + 0.5279z^{-3} + 0.2639z^{-4} + 0.0528z^{-5}) / (1 + 0.6334z^{-2} + 0.0557z^{-4})$
(d) $X(z) = (0.0025z^{-1} + 0.0099z^{-2} + 0.0149z^{-3} + 0.0099z^{-4} + 0.0025z^{-5}) / (1 - 2.9141z^{-1} + 3.5179z^{-2} - 2.0347z^{-3} + 0.4729z^{-4})$

5. Should the roots of a polynomial with real, random coefficients be real or complex? If complex, what special characteristics should they have? Distinguish between polynomials of even length and odd length. An easy way to investigate this is to repeatedly run the code

$$x = roots([randn(1,N)])$$

where N is a conveniently small integer such as 4 or 5. Justify or prove your conclusion.

6. Determine an inverse signal that will result in the unit impulse sequence when processed by the following systems, some of which are unstable.

(a) $y[n] = x[n] + 1.05y[n - 1]$
(b) $y[n] = x[n] - y[n - 2]$
(c) $y[n] = x[n] + 0.5x[n - 1] - 0.25x[n - 2]$

7. For each of the following sets of zeros and poles, which are associated with an underlying causal LTI system, determine the equivalent z-transform (with ROC) and the difference equation. For (c), write m-code to plot the zeros in the z-plane.

(a) zeros = [j,-1,-j]; poles = [0.9*exp(j*pi/20), 0.9*exp(-j*pi/20)]
(b) zeros = [-1,-1,-1]; poles = [0.5774*j, -0.5774*j,0]
(c) zeros = [exp(j*(pi/4:pi/20:3*pi/4)),exp(-j*(pi/4:pi/20:3*pi/4))]

8. A certain sequence $x[n]$ has $X(z) = 1 - z^{-1} + 2z^{-2}.(z \neq 0)$ Determine the z-transforms and corresponding ROCs for the following sequences:

(a) $x_1[n] = 2x[n] - x[n - 1]$

(b) $x_2[n] = 2x[2 - n] - x[n - 1]$
(c) $x_3[n] = nx[1 - n] - x[n - 1]$

9. Using Eq. (2.8), evaluate the magnitude of response of the following systems, specified with poles (pls) and zeros (zzs), to unity-amplitude complex exponentials having the following radian frequencies: $\pi/4, \pi/2. -\pi/4, 0, \pi$.

(a) **zzs = [-1,-1]; pls = [(0.3739 + 0.3639*j), (0.3739 - 0.3639*j)]**
(b) **zzs = [-1,-1,1,1]; pls = [(-0.26 + 0.76*j), (-0.26 - 0.76*j), (0.26 + 0.76*j), (0.26 - 0.76*j)]**

10. Write a script that implements the following call syntax:

$$LVxFreqRespViaZxform(Imp, SR)$$

as described in the text, and which creates a plot of the magnitude and phase of both 1) the z-transform, and 2) the DTFT of Imp.

11. Write a script that implements the function $LVxPlotZXformMagCoeff$ introduced in the text, according to the following format:

function LVxPlotZXformMagCoeff(b,a,rLim,Optionalr0,NSamps)
% intended for use with positive-time z-**transforms, i.e., those**
% whose ROC lies outside a circle of radius equal to the mag-
% nitude of the largest pole; both a and b must be passed in the
% order of: Coeff for z^0, Coeff for z^-1, Coeff for z^-2, etc.
% b **is the vector of numerator coefficients;** a **is the vector of**
% denominator coefficients;
% rLim is the number of circular contours along which to evaluate
% the z-transform. The spacing between multiple contours is 0.02.
% Optionalr0 is the radius of the first contour–if passed as [], the
% default is 1.0 for FIRs (i.e., DenomCoeffVec=[1]) or the magni-
% tude of the largest pole plus 0.02. If the largest pole magnitude
% is > 0.98 and < 1.0, the radius of the first contour is set
% at 1.0 unless Optionalr0 is larger than 1.0. If there are no poles,
% pass a **as [1]. If there are no zeros, pass** b **as [1].**
% NSamps is the number of frequencies at which to evaluate the
% z-**transform along each contour.**
% Sample calls:
% LVxPlotZXformMagCoeff([1],[1,-1.8,0.81],[1],[1],256)
% LVxPlotZXformMagCoeff([1],[1,0,0.81],[1],[1],256)
% LVxPlotZXformMagCoeff([1,1,1,1,1,1,1,1,1],[1],[1],[1],256)

12. Write a script that implements the function $LVxPlotZTransformMag$ as described in the text and further defined below. The script should plot at least the magnitude of the z-transform of the transfer function defined by the input arguments $ZeroVec$ and $PoleVec$.

```
function LVxPlotZTransformMag(ZVec,PVec,rMin,rMax,NSamps)
% LVxPlotZTransformMag is intended for use with right-handed
% z-transforms, i.e., those whose ROC is the area outside of a
% circle having a radius just larger than the magnitude of the
% largest pole of the z-transform. ZVec is a row vector of zeros;
% if no zeros, pass as []
% PVec is a row vector of poles; if no poles, pass as []
% rMin and rMax are the radii of the smallest and largest contours
% to compute; rMin and rMax may be equal to compute one con-
% tour. Pass rMin = rMax = 1 for standard frequency response.
% NSamps is the number of frequencies at which to evaluate the
% z-transform along each contour. rMin is set at 0.02 greater than
%  the magnitude of the largest pole unless the pole magnitude is
% greater than 0.98 and less than 1.0 in which case rMin is set
% at 1.0.
% Sample calls:
% LVxPlotZTransformMag([-1,1],[0.95*exp(j*2*pi/5),...
% 0.95*exp(-j*2*pi/5)],1,1,256)
% LVxPlotZTransformMag([exp(j*2*pi/4) exp(-j*2*pi/4)],...
% [0.95*exp(j*2*pi/4),0.95*exp(-j*2*pi/4)],1,1,256)
% LVxPlotZTransformMag([-1,j,-j],[],1,1,256)
% LVxPlotZTransformMag([],[0.9],1,1,256)
% LVxPlotZTransformMag([(0.707*(1+j)),j,(0.707*(-1+j)),...
% -1,(0.707*(1-j)),-j,-(0.707*(1+j))],[],1,1,256)
```

13. Evaluate the frequency response of the z-transforms below at radian frequencies 0, $\pi/4$, and $\pi/2$, then obtain the impulse response for the first 150 samples and correlate it with the complex exponential corresponding to the same radian frequencies. Plot the impulse response and the real and imaginary parts of the exponential power series used in the correlation. Compare the correlation values to the answers obtained directly from the z-transform. The z-transforms are

a) $X(z) = [1 + z^{-1}]/(1 - 0.9z^{-1})$; ROC: $|z| > 0.9$

b) $X(z) = [z^{-1} + z^{-2}]/(1 - z^{-1} + 0.9z^{-2})$; ROC: $|z| > 0.9487$

c) $X(z) = [1 + z^{-4}]/(1 - 1.3z^{-1} + 0.95z^{-2})$; ROC: $|z| > 0.9747$

14. Write a script that can receive set of b and a coefficients of the z-transform of a causal sequence whose ROC either lies outside a circle of radius equal the magnitude of the largest pole, or includes the entire complex plane except $z = 0$, and compute the time domain sequence $x[n]$ using the method of partial fraction expansion, employing the function *residuez*. Test your script on the following z-transform coefficients:

(a) b = [0.527,0,1.582,0,1.582,0,0.527]; a = [1,0,1.76, 0,1.182, 0,0.278]

(b) b = [0.518 0 1.554 0 1.554 0 0.518]; a = [1,0,1.745,0,1.172,0,0.228]

(c) **[b,a] = butter(8,[0.4,0.6])**
(d) **[b,a] = cheby1(8,1,[0.6,0.8],'stop')**

15. Determine the frequency response of the following LTI systems, characterized by their poles and zeros, at each of the following radian frequencies: $0, \pi/4, \pi/2, 3\pi/4, \pi$.

(a) Poles = [0.9,0.9,0.9j,-0.9j]; Zeros = [0.65(1+j),0.65(1-j),-0.65(1+j),-0.65(1-j)]
(b) Poles = [(0.0054 ± 0.9137j),(0.3007 ± 0.6449j),0.551]; Zeros = [-1,-1,-1,-1,1]
(c) Poles = [(-0.5188 ± 0.7191j), ±0.4363,(0.5188 ± 0.7191j)]; Zeros = [±j, ±j, ±j]
(d) Poles = [(±0.5774j),0]; Zeros = [1,1,1]

16. Write a script that implements a numerical approximation of Eq. (2.17) according to the following function specification:

function LVxNumInvZxform(b,a,M,ContourRad,nVals)
%
% Intended to perform the inverse z-transform for transforms whose
% ROC lies outside a circle of radius equal to the largest pole
% magnitude of the z-transform b is the numerator coefficient vector;
% a is the denominator coefficient vector
% M is the number of samples to use along the contour
% ContourRad is the radius of the circle used as the contour
% nVals is vector of time domain indices to compute, i.e., x[nVals]
% will be computed via the inverse z-transform
% Sample Call:
% **LVxNumInvZxform([1],[1,-0.9],5000,1,[0:1:20])**
% **LVxNumInvZxform([0 1],[1,-2,1],5000,1.1,[0:1:20])**
% **LVxNumInvZxform([1],[1,-1.2,0.81],60000,0.91,[0:1:100])**
% **LVxNumInvZxform([1],[1,-1.2,0.81],5000,1,[0:1:100])**

A simple way to perform numerical integration of, for example, the function $y = x^2$, is to construct small boxes of width Δx and height $f(x)$ under the curve and take the limit of the sum of the areas of the boxes as $\Delta x \to 0$. Figure 2.36, plot (a) shows this approach. An approach that uses trapezoids instead of rectangles, and which converges more quickly, is shown in plot (b). It is easy to show that the area of each trapezoid is Δx multiplied by

$$0.5(f(x[n]) + f(x[n+1]))$$

where

$$\Delta x = x[n+1] - x[n]$$

This latter method, the trapezoidal, is the one you should use for your script as the faster convergence greatly reduces the number of sample points needed for acceptable results.

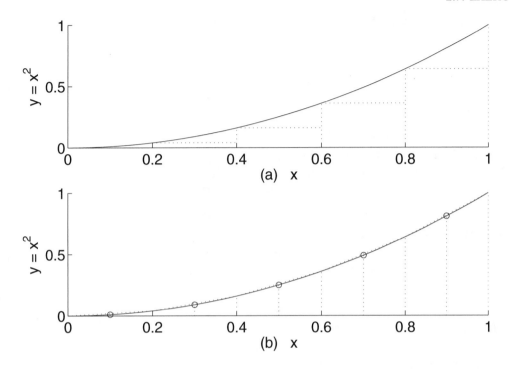

Figure 2.36: (a) Rectangular approach to numerically estimating the integral of the function $y = x^2$ over the interval 0:1; (b) Trapezoidal approach to numerically estimating the integral of the function $y = x^2$ over the interval 0:1, with the average function value on the interval marked at the mid-point of each interval with a circle.

17. Write a script that will receive the location of a zero of the form $M \exp(j\theta)$, where M and θ are real numbers, construct a z-transform consisting of the following four zeros: $M \exp(j\theta)$, $M \exp(-j\theta)$, $(1/M) \exp(j\theta)$, and $(1/M) \exp(-j\theta)$, two different ways.

   First, using paper and pencil, derive an algebraic expression in M and θ for the z-transform coefficients, and second, construct the z-transform coefficients using m-code to generate the zero locations and from them generate the polynomial coefficients (using, for example, the function *poly*) for the z-transform.

   Next, using the values of z given below, compute the z-transform coefficients using both methods, i.e., using your algebraic expression to compute the coefficients, and second, using the m-code method to obtain the zeros and compute the z-transform coefficients using *poly*. The two results should agree.

   Next, evaluate the z-transform along the unit circle at 256 points from $-\pi$ to π, and plot the magnitude and phase of the z-transform. Test your script with the following zeros, and comment on the phase characteristics of the LTI system defined by the z-transform constructed from four zeros as specified above.

(a) $0.6 + 0.6j$

(b) $0.95j$

(c) -0.9

(d) $(\sqrt{2}/2)(1 + j)$

18. Write the m-code for the script *LVxDir2Parallel* as specified below:

    function [Bp,Ap,Cp] = LVxDir2Parallel(b,a)
    % Receives a set of Direct Form coefficients b and a and
    % computes the Parallel Form coefficients Bp, Ap, and Cp

19. Write the m-code for the script *LVxFilterParallelForm* as specified below:

    function [y] = LVxFilterParallelForm(Bp,Ap,Cp,x)
    % Receives a set of Parallel Form coefficients and a sequence x
    % and filters x in a Parallel Form structure, delivering the filtered
    % output as y

20. Write the m-code for the script *LVxParallel2Dir* as specified below:

    function [b,a] = LVxParallel2Dir(Bp,Ap,Cp)
    % Receives a set of Parallel Form coefficients Bp, Ap,
    % and Cp, and converts them to a set of Direct Form
    % coefficients [b,a]

21. For the following LTI systems, sketch and label with coefficients the Direct Form I, Direct Form II, Cascade (using Direct Form II sections), Parallel (using Direct Form II sections), and Lattice implementations.

    (a) $y[n] = \sum_{m=0}^{2}(1/3)^m + \sum_{n=1}^{3}(1/2)^n$

    (b) **b = [0.0219, 0.1097, 0.2194, 0.2194, 0.1097, 0.0219], a = [1, -0.9853, 0.9738, -0.3864, 0.1112, -0.0113]**

    (c) **b = [0.0109, 0.0435, 0.0653, 0.0435, 0.0109], a = [1, -2.1861, 2.3734, -1.3301, 0.3273]**

22. Convert the following sets of Cascade Form coefficients into Direct Form coefficients, and then into Parallel Form coefficients. Sketch the Cascade, Direct, and Parallel filter signal flow diagrams. To verify equivalence, filter a linear chirp using all three filter forms, and plot and compare the results from each filtering operation.

    (a) **Bc = [1, 2.002, 1.002; 1, 1.9992, 0.9992; 0, 1, 0.9987]; Ac = [1, 0.5884, 0.8115; 1, -0.0714, 0.4045; 0,1, -0.3345]; Gain = 0.0665**

    (b) **Bc = [1,-2.0001, 1.0001; 1,-1.9999, 0.9999; 1,2,1; 1,2,1]; Ac = [1, 0.5973, 0.9023; 1, -0.5973, 0.9023; 1, 0.2410, 0.7621; 1, -0.2410, 0.7621]; Gain = 0.0025**

23. Convert the following Parallel Form filters into equivalent Lattice-Ladder filters, filter a linear chirp using both filters, and plot and compare the results from each filtering operation to verify equivalence.

(a) **Bp = [-0.0263,-0.0846;-0.0263,0.0846;-0.6919,0.4470; -0.6919,-0.4470]; Ap = [1,-0.5661,0.9068; 1,0.5661,0.9068;1,-0.5713,0.5035; 1,0.5713,0.5035]; Cp = 1.8148**

(b) **Bp = [-0.3217,-0.0475; -0.3217,0.0475; -0.7265,0]; Ap = [1,-0.4764,0.7387; 1,0.4764,0.7387; 1,0, 0.5095]; Cp = 1.8975**

24. Assume that an impulse response that is symmetrical about its midpoint has a linear phase characteristic. Prove that the following LTI systems have linear phase characteristics:

(a) A system having only single real zeros of magnitude 1.0 (i.e., zeros equal only to 1 or -1).

(b) A system having pairs of zeros that are real only and in reciprocal-magnitude pairs, i.e., of the form $[M, (1/M)]$ where M is real.

(c) A system having pairs of zeros that are complex conjugate pairs having magnitude 1.0.

(d) A system having quads of zeros, i.e., a set of four zeros of the form

$$[M \angle \theta, M \angle(-\theta), (1/M) \angle \theta, (1/M) \angle(-\theta)]$$

25. This exercise will copy a VI and modify m-code in its MathScript node to change its behavior for comparison to the original version. The VI *DemoDragZerosZxformVI* constructs an FIR from, in general, a quad of four zeros when in quad mode. However, it uses only one or two zeros when certain conditions occur such that only the one zero or pair of zeros are necessary to create a linear phase FIR (see discussion in this chapter). In this exercise, we'll copy the VI to a new file name and modify it to use four zeros in all situations to see what the difference is in transfer functions for the same main zero location.

Open the VI *DemoDragZerosZxformVI* and use *Save As* from the *File* menu to copy the original VI to a new file name. In the new VI, go to the Window menu and select *Show Block Diagram* (or Press Control+E). You should see a large box in the Block Diagram filled with m-code; modify this code appropriately to use all four zeros for all situations, save the new VI with your m-code modifications, and run the new VI in quad mode (four zeros). If you have done your work properly, all four zeros will be used for all situations. When this is the case, the impulse response will always be five samples in length. Compare impulse response length and magnitude and phase responses to those obtained with the original VI for the following values of the main zero:

(a) $z = 1.0$

(b) $z = -1.0$

(c) $z = 0.707 + 0.707j$

CHAPTER 3

The DFT

3.1 OVERVIEW

In previous chapters, we have studied the DTFT and the z-transform, both of which have important places in discrete signal processing theory, but neither is a numerically computable transform. Signal processing as applied in industry and commerce, however, for such applications as audio and video compression algorithms, etc., relies on computable transforms. Unlike the DTFT and z-transform, however, the Discrete Fourier Transform is a numerically computable transform. It is a reversible frequency transform that evaluates the spectrum of a finite sequence at a finite number of frequencies. It is perhaps the best known and most widely used transform in digital signal processing. Many papers and books have been written about it. There are other numerically computable frequency transforms, such as the Discrete Cosine Transform (DCT), the Modified Discrete Cosine Transform (MDCT), the Discrete Sine Transform (DST), and the Discrete Hartley Transform (DHT), but the DFT is the most often discussed and used frequency transform. Its fast implementation, the FFT (actually an entire family of algorithms) serves as the computational basis not only for the DFT per se, but for other transforms that are related to the DFT but do not have their own fast implementations. While the end goal of much DFT use may be spectral analysis, there is a growing body of applications that use the DFT-IDFT (and relatives of it, such as the Discrete Cosine Transform) for data compression.

In this chapter we cover the DFS and its basis, the reconstruction of a sequence from samples of its z-transform, followed by the definition(s), basic properties, and computation of the DFT. We then delve into a mix of practical and theoretical matters, including the DFTs of common signals, determination of Frequency Resolution/Binwidth, the FFT or Fast Fourier Transform, the Goertzel Algorithm (a simple recursive method to compute a single DFT bin which finds utility in such things as DTMF detection and the like), implementation of linear convolution using the DFT (a method permitting efficient convolution of large blocks of data), DFT Leakage (an issue in spectral analysis and signal detection), computation of the DTFT via the DFT, and the Inverse DFT (IDFT), which we compute directly, by matrix methods, or through ingenious use of the DFT.

3.2 SOFTWARE FOR USE WITH THIS BOOK

The software files needed for use with this book (consisting of m-code (.m) files, VI files (.vi), and related support files) are available for download from the following website:

http://www.morganclaypool.com/page/isen

The entire software package should be stored in a single folder on the user's computer, and the full file name of the folder must be placed on the MATLAB or LabVIEW search path in accordance with the instructions provided by the respective software vendor (in case you have encountered this notice before, which is repeated for convenience in each chapter of the book, the software download only needs to be done once, as files for the entire series of four volumes are all contained in the one downloadable folder).

See Appendix A for more information.

3.3   DISCRETE FOURIER SERIES

From our earlier introductory discussion, several chapters ago, recall that the DFS coefficients $\widetilde{X}[k]$ of a periodic sequence $x[n]$ $(-\infty < n < \infty)$ are

$$DFS(x[n]) = \widetilde{X}[k] = \sum_{n=0}^{N-1} \widetilde{x}[n]e^{-j2\pi nk/N} \tag{3.1}$$

where k is an integer and $\widetilde{x}[n]$ is one period of the periodic sequence $x[n]$. Typical ranges for k are: $0:1:N$-1, or $-N/2+1:1:N/2$ for even length sequences, or $-(N$-1$)/2:1:(N$-1$)/2$ for odd length sequences. The sequence $x[n]$ can be reconstructed from the DFS coefficients as follows:

$$x[n] = \frac{1}{N} \sum_{k=0}^{N-1} \widetilde{X}[k]e^{j2\pi nk/N} \tag{3.2}$$

Note that not only is $x[n]$ periodic over n, but the DFS coefficients $\widetilde{X}[k]$ are periodic over k.

Example 3.1.   One period of a certain periodic sequence is $[(2 + j),$-1, $j, 3]$; compute the DFS coefficients, and then reconstruct one period of the sequence $x[n]$ using the coefficients.

A straightforward script using a single loop with n as a vector would be

```
x = [(2+j),-1,j,3]; N = length(x);
n = 0:1:N-1;  W = exp(-j*2*pi/N);
for k = 0:1:N-1, DFS(k +1) = sum(x.*(W.^(n*k))); end
```

And reconstruction can be performed with this code:

```
N = length(DFS); k = 0:1:N-1; W = exp(j*2*pi/N);
for n = 0:1:N-1, x(n +1) = sum(DFS.*(W.^(n*k))); end
x = x/N
```

More efficient (vectorized) code to perform the analysis and synthesis, respectively, is

```
x = [(2+j),-1,j,3]; N = length(x); n = 0:1:N-1;
k = n; DFS = x*(exp(-j*2*pi*(n'*k)/N))
```

```
N = length(DFS); n = 0:1:N-1;
```

k = n; x = DFS*(exp(j*2*pi*(n*k)/N))/N

Example 3.2. Evaluate the DFS of the sequence [1, 0, 1] and plot and discuss the result, contrasting it to the DTFT of the sequence.

The following code will work for any sequence x:

```
x = [1,0,1]; N = length(x); n = 0:1:N-1; W = exp(-j*2*pi/N);
k = 0:1:N-1; DFS = x*(W.^(n*k)),
figure(9); stem(n,abs(DFS))
xlabel('Normalized Frequency'); ylabel('Magnitude')
```

Observing the magnitude plot, it is difficult to ascertain the true frequency content of the sequence with only three frequency points. We can determine, however, where the three DFS frequencies lie on the DTFT plot. The three DFS frequencies are $2\pi k/3$ where $k = 0,1$, and 2. To get both the DTFT and the three DFS samples onto the same plot, the following code evaluates the DTFT from 0 to 2π radians, allowing all three DFS frequencies to be plotted:

```
x = [1 0 1]; N = length(x); n = 0:1:N-1; k = n;
W = exp(-j*2*pi/N); DFS = x*(W.^(n*k));
w = 0:0.01:2*pi; DTFT = 1+exp(-j*2*w);
figure(8); clf; hold on; plot(w/(pi),abs(DTFT));
stem(2*[0:1:2]/3,abs(DFS),'ro');
xlabel('Norm Freq (Units of pi)'); ylabel('Magnitude')
```

Figure 3.1 illustrates the essential results from the code above.

From the above it is possible to see that the DFS coefficients of a sequence $x[n]$ (computed using one period $\tilde{x}[n]$ of $x[n]$) are essentially equally spaced samples of the DTFT of $\tilde{x}[n]$. Assuming that the ROC of the z-transform of $\tilde{x}[n]$ includes the unit circle, then it is also possible to say that the DFS coefficients are essentially equally spaced samples of the z-transform of $\tilde{x}[n]$ evaluated along the unit circle. Prior to proceeding to the DFT, we briefly investigate the effect of sampling the z-transform.

3.4   SAMPLING IN THE Z-DOMAIN

The z-transform of an absolutely summable sequence $x[n]$, which may be finite or infinite in extent, is defined as

$$X(z) = \sum_{n=-\infty}^{\infty} x[n]z^{-n}$$

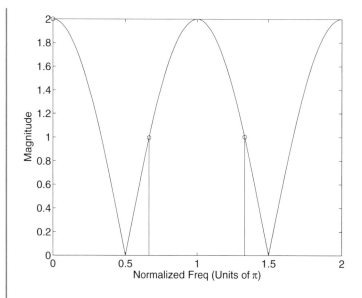

Figure 3.1: A dense grid of samples of the DTFT of the sequence [1 0 1], with a stem plot of the three DFS samples.

Assuming that the ROC includes the unit circle, and sampling $X(z)$ at highfrequencies whose radian arguments are $2\pi k/M$ where M is an arbitrary integer and $k = 0, \pm 1, \pm 2 ...$we define

$$\widetilde{X}[k] = X(z)|_{z=\exp(j2\pi k/M)} = \sum_{n=-\infty}^{\infty} x[n]e^{-j2\pi kn/M} \tag{3.3}$$

which is periodic over M, and has the form of a Discrete Fourier Series. Computing the inverse DFS of $\widetilde{X}[k]$ will result in a periodic sequence $\widetilde{x}[n]$ where

$$\widetilde{x}[n] = \sum_{p=-\infty}^{\infty} x[n - pM] \tag{3.4}$$

When the number of samples M taken of $X(z)$ is smaller than N, the original length of $x[n]$, which may be infinite or finite, aliasing occurs. If M samples of $X(z)$ are obtained, the reconstructed sequence $\widetilde{x}[n]$ consists of a superposition of copies of $x[n]$ offset in time by multiples of M samples according to Eq. (3.4).

Example 3.3.   Write a script that can receive a set of coefficients $[b, a]$ representing the z-transform (the ROC of which includes the unit circle) of a time domain sequence $x[n]$, a desired number *NumzSamps* of samples of the z-transform, and a desired number of time domain samples $nVals$ to reconstruct, and reconstruct $x[n]$ to demonstrate the periodicity of the reconstruction. Use $x[n]$

= $u[n] - u[n - 4]$ and test the reconstruction for values of *NumzSamps* = 8, 5, and 3, and plot the results. Compute several periods of $\tilde{x}[n]$ as given by Eq. (3.4) and compare results to those obtained using the script below.

The script is straightforward, and the results are shown in Fig. 3.2.

```
function LV_TDReconViaSampZXform(b,a,NumzSamps,n)
kz = 0:1:NumzSamps-1;
z = exp(j*2*pi*(kz/NumzSamps));
Nm = 0; Dm = 0;
for nn = 0:-1:-length(b)+1
 Nm = Nm + (b(-nn+1))*(z.^nn); end
for d = 0:-1:-length(a)+1
 Dm = Dm + (a(-d+1))*(z.^d); end
zSamps = Nm./Dm;
figure(125);
clf; N = length(zSamps);
W = exp(j*2*pi/N); k = 0:1:N-1;
IDFS = real((W.^((n')*k))*conj(zSamps')/N);
stem(n,IDFS)
% Test calls:
% LV_TDReconViaSampZXform([ones(1,4)],[1],8,[-15:1:15])
% LV_TDReconViaSampZXform([ones(1,4)],[1],5,[-15:1:15])
% LV_TDReconViaSampZXform([ones(1,4)],[1],3,[-15:1:15])
```

To use Eq. (3.4), we can sum a few offset versions of $x[n]$ to obtain several periods of the periodic sequence. For the case of $M = 3$, we can sum $x[n]$, $x[n - 3]$, $x[n + 3]$, $x[n - 6]$, $x[n + 6]$, etc. This can be done by repetitively using the script $[y, nOut] = LVAddSeqs$ which was introduced in Volume I of the series (see the Preface to this volume for the contents of Volume I of the series). The results, shown in Fig. 3.3, are valid from n = -3 to n = +6

```
x = [1,1,1,1]; n = [0,1,2,3];
[y, nOut] = LVAddSeqs(x,n,x,n+3);
[y, nOut] = LVAddSeqs(y,nOut,x,n-3);
[y, nOut] = LVAddSeqs(y,nOut,x,n+6);
[y, nOut] = LVAddSeqs(y,nOut,x,n-6);
figure(7); stem(nOut,y)
```

We have just seen that a reconstruction of $x[n]$ using samples of $X(z)$ results in a periodic version of $x[n]$. It is also possible, under certain conditions given below, to completely reconstruct $X(z)$ from $x[n]$, which implies that the original $x[n]$ (i.e., not a periodic version of it) can be reconstructed by using the inverse z-transform on the reconstructed $X(z)$.

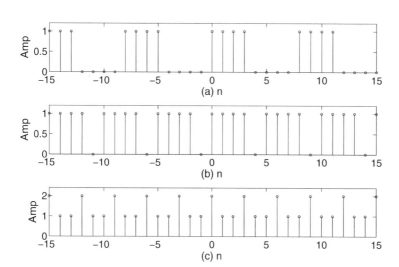

Figure 3.2: (a) Periodic reconstruction of $x[n]$ ($= u[n]$-$u[n-4]$) from eight samples of $X(z)$; (b) Periodic reconstruction from five samples of $X(z)$; (c) Periodic reconstruction from three samples of $X(z)$.

The conditions under which $X(z)$ can be completely reconstructed are as follows: if a time domain sequence $x[n]$ is of finite extent, i.e., $x[n]$ is identically zero for $n < 0$ and $n \geq N$, and the unit circle is in the ROC, then $\widetilde{X}[k]$, which consists of N samples of $X(z)$ along the unit circle, located at frequencies specified as $2\pi k/N$, where $k = 0{:}1{:}N-1$, can determine $X(z)$ for all z according to the following equation:

$$X(z) = \frac{1-z^{-N}}{N} \sum_{k=0}^{N-1} \frac{\widetilde{X}[k]}{1 - e^{j2\pi k/N} z^{-1}} \tag{3.5}$$

Example 3.4.   For the sequence [1, 1, 1, 1], compute and display 1000 samples of the magnitude of the z-transform directly, then compute and display 1000 samples of the magnitude of the z-transform using Eq. (3.5).

In the m-code below, we first compute 1000 points of the z-transform by writing the z-transform of the sequence and evaluating; we then extract four properly located samples from the 1000 samples and use Eq. (3.5) to reconstruct the entire 1000 samples of the z-transform. The results are shown in Fig. 3.4.

```
inc = 1/999; xvec = inc/2:inc:1; zp = 2*pi*xvec; z = exp(j*zp);
ZX1 = 1 + z.^(-1) + z.^(-2) + z.^(-3);
```

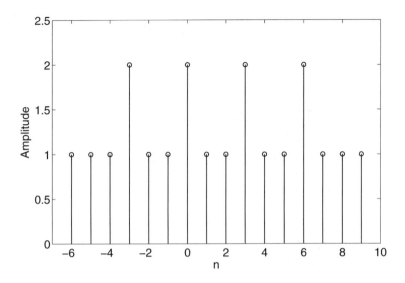

Figure 3.3: A partial reconstruction (using Eq. (3.4)) of the infinite-extent periodic sequence $\tilde{x}[n]$ that results from reconstruction of $x[n]$ from samples of the z-transform $X(z)$, where $x[n] = [1,1,1,1]$.

```
figure(33); subplot(211);
plot(xvec,abs(ZX1)); axis([0,1,0,5])
k = 0:1:3; Fndx = fix((k/4)*1000) + 1;
Xtil = ZX1(Fndx); S = 0;
for k = 0:1:3
S = S + Xtil(k+1)./(1-(exp(j*2*pi*k/4))*(z.^(-1)));
end
ZX2 = S.*((1-z.^(-4))/4);
subplot(212); plot(xvec,abs(ZX2))
axis([0,1,0,5])
```
The script

$$LVxZxform From Samps(b, a, Numz Samps, M, nVals)$$

will, for a set of coefficients $[b, a]$, compute samples of $X(z)$, then reconstruct $X(z)$, then reconstruct $x[n]$ using contour integration, and, for contrast, reconstruct a periodic version of $x[n]$ using the inverse DFS transform on the samples of $X(z)$. The call

LVxZxformFromSamps([1,1,1,1],[1],6,5000,30)

results in Fig. 3.5.

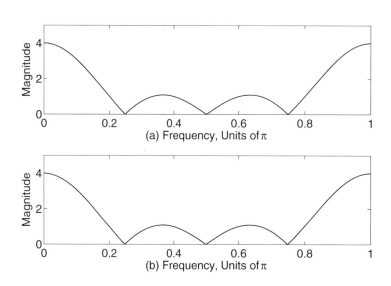

Figure 3.4: (a) 1000 samples of the z-transform $X(z)$ of the sequence $[1,1,1,1]$, obtained by direct evaluation of the z-transform $X(z) = 1 + z^{-1} + z^{-2} + z^{-3}$;(b) 1000 samples of the z-transform of the same sequence, reconstructed from four samples of $X(z)$ using Eq. (3.5).

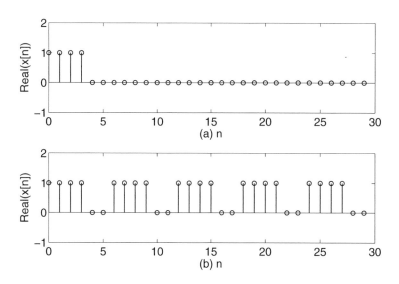

Figure 3.5: (a) A reconstruction of $x[n] = [1,1,1,1]$ from a reconstruction of $X(z)$ based on six samples of $X(z)$; (b) A reconstruction of a periodic version of $x[n]$ directly from the six samples of $X(z)$.

3.5 FROM DFS TO DFT

In the DFS, we have a computable transform; a periodic sequence $x[n]$ of periodicity N can be reconstructed from N (correlation) coefficients $\widetilde{X}[k]$ obtained from N correlations performed between N harmonically related basis functions and a single period of $x[n]$, designated $\widetilde{x}[n]$.

In many cases, however, a sequence $y_1[n]$ ($0 \le n \le N - 1$) to be analyzed is not periodic, but rather aperiodic and finite. The DFT is defined for such sequences by stipulating that $y_1[n]$ is hypothetically a single period $\widetilde{y}[n]$ of an infinite length periodic sequence $y[n]$. By entertaining this line of argument, it is possible to decompose a finite sequence of length N into N coefficients, and reconstruct it using essentially the same basis functions for analysis and synthesis used for the DFS.

Thus, the DFS analysis formula

$$DFS(x[n]) = \widetilde{X}[k] = \sum_{n=0}^{N-1} \widetilde{x}[n]e^{-j2\pi nk/N}$$

can be rewritten using notation for a finite sequence as

$$DFT(x[n]) = X[k] = \sum_{n=0}^{N-1} x[n]e^{-j2\pi nk/N}$$

and the DFS synthesis formula

$$x[n] = \frac{1}{N} \sum_{k=0}^{N-1} \widetilde{X}[k]e^{j2\pi nk/N}$$

can be rewritten as

$$x[n] = \frac{1}{N} \sum_{k=0}^{N-1} X[k]e^{j2\pi nk/N}$$

3.6 DFT-IDFT PAIR

3.6.1 DEFINITION-FORWARD TRANSFORM (TIME TO FREQUENCY)

The complex Discrete Fourier Transform (DFT) may be defined as:

$$X[k] = \sum_{n=0}^{N-1} x[n](\cos[2\pi nk/N] - j\sin[2\pi nk/N]) \tag{3.6}$$

or, using the complex exponential form,

$$X[k] = \sum_{n=0}^{N-1} x[n]e^{-j2\pi nk/N} \tag{3.7}$$

where n is the sample index (running from 0 to $N-1$) of an N point sequence $x[n]$. Harmonic number (also called mode, frequency, or bin) is indexed by k and runs from 0 to $N-1$.

Using Eq. (3.7) results in computation by the **Direct Method**, as opposed to much more efficient algorithms lumped under the umbrella term **FFT (Fast Fourier Transform)**.

3.6.2   DEFINITION-INVERSE TRANSFORM (FREQUENCY TO TIME)

If the DFT is defined as in Eq. (3.6), then the Inverse **Discrete Fourier Transform (IDFT)** is defined as:

$$x[n] = \frac{1}{N} \sum_{k=0}^{N-1} X[k](\cos[2\pi kn/N] + j \sin[2\pi nk/N]) \tag{3.8}$$

for $n = 0$ to $N-1$, or, in complex exponential notation

$$x[n] = \frac{1}{N} \sum_{k=0}^{N-1} X[k]e^{j2\pi nk/N} \tag{3.9}$$

3.6.3   MAGNITUDE AND PHASE

Since DFT bins are, in general, complex numbers, the magnitude and phase are computed as for any complex number, i.e.,

$$|X[k]| = \sqrt{\mathrm{Re}(X[k])^2 + \mathrm{Im}(X[k])^2}$$

and

$$\angle X[k] = \mathrm{arcTan}(\mathrm{Im}(X[k])/\mathrm{Re}(X[k]))$$

3.6.4   N, SCALING CONSTANT, AND DFT VARIANTS

The definitions above are independent of whether N is even or odd. An alternate definition of the DFT for N even makes the ranges of n and k run from $-N/2 + 1$ to $N/2$. An alternate definition of the DFT for N odd has the ranges of n and k running from $-(N-1)/2$ to $(N-1)/2$.

Note in Eqs. (3.8) and (3.9) above that the scaling constant $1/N$ was applied to the Inverse DFT, and no such constant was applied to the DFT. In some DFT definitions, the scaling constant $1/N$ is applied to the DFT rather than to the Inverse DFT. It is also possible to apply a scaling constant of $1/\sqrt{N}$ to both the DFT and IDFT.

There are yet other ways of defining the DFT; they all result in essentially the same information, however. The variations consist of different ways of defining the range of k or n, and whether or not the DFT or the IDFT has a scaling coefficient (such as $1/N$) applied to it. Reference [1], which is devoted entirely to the DFT, discusses a number of these variations.

3.7   MATHSCRIPT IMPLEMENTATION

Since MathScript vectors cannot use index values that are equal to or less than zero, MathScript implements the complex DFT in this way:

$$X[k] = \sum_{n=1}^{N} x[n]e^{-j2\pi(k-1)(n-1)/N} \tag{3.10}$$

where both n and k run from 1 to N. Note that the MathScript version does not scale the DFT by $1/N$–instead, the factor $1/N$ is applied in the function $ifft$, which is MathScript's inverse DFT function. When using the standard DFT definition Eq. (3.7), since the factor $1/N$ is applied in the DFT itself, it is not applied when using the IDFT.

MathScript can be used to compute the DFT of a sequence by using the function *fft*.

Example 3.5.   Compute and display the magnitude and phase of the DFT of the signal sequence $[ones(1, 64)]$ using MathScript.

We make the call

$$s = [ones(1,64)]; y = fft(s);$$

and follow it with

$$subplot(2,1,1); stem(abs(y)); subplot(2,1,2); stem(unwrap(angle(y)))$$

3.8   A FEW DFT PROPERTIES

Before proceeding, it should be noted that a number of DFT properties pertain to sequences or their DFTs that have been circularly folded. A circular folding of a sequence $x[n]$ of length N is defined as

$$x[(-n)]_N = \begin{cases} x[0] & n = 0 \\ x[N-n] & 1 \leq n \leq N-1 \end{cases}$$

1. Linearity
The DFT is a linear operator, i.e., the following is true:

$$DFT(ax_1[n] + bx_2[n]) = aDFT(x_1[n]) + bDFT(x_2[n])$$

2. Circular folding
If the DFT of a sequence $x[n]$ is $X[k]$, the DFT of a circularly folded version of $x[n]$ is a circularly folded version of the DFT of the sequence, i.e.,

$$DFT(x[(-n)]_N) = X[(-k)]_N$$

The following code demonstrates this property.

```
n = 0:1:7; x = [0:1:7]; xret = x(mod(-n,8)+1);
subplot(3,2,1);stem(n,x); y = fft(x); yret = fft(xret);
subplot(3,2,2);stem(n,xret); subplot(3,2,3); stem(n,real(y));
subplot(3,2,4);stem(n,real(yret));subplot(3,2,5); stem(n,imag(y));
subplot(3,2,6); stem(n,imag(yret))
```

3. Shift (circular) in Time Domain

Since the DFT is periodic in n, shifting the sequence some number of samples m to the right can be equivalently achieved by a circular shift, and

$$DFT(x[n - m]_N) = X(k)e^{-j2\pi mk/N}$$

The following code verifies this for a short sequence. The DFT *prxshfft* is constructed according to the formula from xft, the DFT of the original sequence x.

```
x=[1 2 3 4]; xsh1 = [4 1 2 3]; xft = fft(x), xshfft = fft(xsh1),
k=0:1:length(x)-1; prxshfft = xft.*(exp(-j*2*pi*1*k/4))
```

4. Circular Convolution of Two Time Domain Sequences

If the circular convolution of two sequences $x_1[n]$ and $x_2[n]$ is defined as

$$x_1[n] \circledast x_2[n] = \sum_{m=0}^{N-1} x_1[m]x_2[(n - m)]_N$$

for $0 \le n \le N - 1$, then

$$DFT(x_1[n] \circledast x_2[n]) = X_1[k]X_2[k]$$

This property and its usefulness to perform ordinary (or linear) convolution will be discussed in detail later in the chapter.

5. Multiplication of Time Domain Sequences

The DFT of the product of two time domain sequences is $1/N$ times the circular convolution of the DFTs of each:

$$DFT(x_1[n]x_2[n]) = \frac{1}{N}(X_1[k] \circledast X_2[k])$$

6. Parseval's Relation

The energy of a sequence is the sum of the squares of the absolute values of the samples, and this has an equivalent computation using DFT coefficients:

$$\sum_{n=0}^{N-1} |x[n]|^2 = \frac{1}{N} \sum_{k=0}^{N-1} |X[k]|^2$$

The following code verifies this property using a random sequence:

N = 9; x = randn(1,N); td = sum(abs(x).^2),
fd = (1/N)*sum(abs(fft(x)).^2)

7. Conjugate Symmetry

The DFT of a real input sequence is **Conjugate-Symmetric**, which means that

$$\text{Re}(X[k]) = \text{Re}(X[-k])$$

and

$$\text{Im}(X[k]) = -\text{Im}(X[-k])$$

This property implies that (for a real sequence $x[n]$) it is only necessary to compute the DFT for a limited number of bins, namely

$$k = 0{:}1{:}N/2 \qquad N \text{ even}$$
$$k = 0{:}1{:}(N-1)/2 \qquad N \text{ odd}$$

Example 3.6.   Demonstrate conjugate symmetry for several input sequences.

We'll use the input sequence $x = [16{:}{-}1{:}{-}15]$. Since the length is even, there are two bins that will not have complex conjugates, namely Bins 0 and 16 in this case, or Bins 0 and $N/2$ generally. The following code computes and displays the DFT; conjugate symmetry is shown except for the aforementioned Bins 0 and 16.

n=[0:1:31]; x=[16:-1:-15]; y=fft(x); figure
subplot(2,1,1); stem(n,real(y)); subplot(2,1,2); stem(n,imag(y))

Figure 3.6 shows the result of the above code.

If we modify the sequence length to be odd, then there is no Bin $N/2$, so conjugate symmetry is shown for all bins except Bin 0. The result from running the code below is shown in Fig. 3.7.

n=[0:1:32]; x=[16:-1:-16]; y=fft(x); figure
subplot(2,1,1); stem(n,real(y)); subplot(2,1,2); stem(n,imag(y))

8. Even/Odd TD-Real/Imaginary DFT Parts

The circular even and odd decompositions of a time domain sequence are defined as

$$x_{cE}[n] = \begin{cases} x[0] & n = 0 \\ (x[n] + x[N-n])/2 & 1 \le n \le N-1 \end{cases}$$

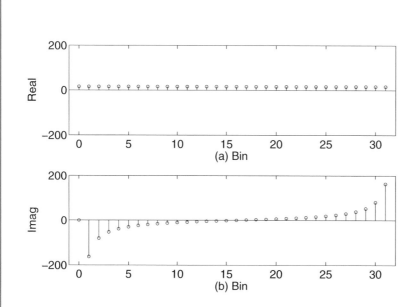

Figure 3.6: (a) Real part of DFT of the sequence $x[n]$ = [16:-1:-15]; (b) Imaginary part of same.

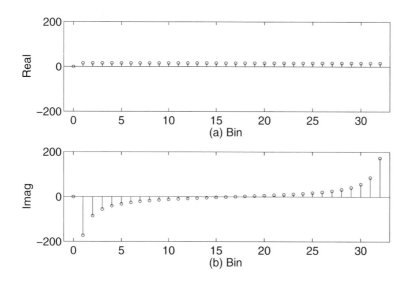

Figure 3.7: (a) Real part of DFT of the sequence $x[n]$ = [16:-1:-16]; (b) Imaginary part of same.

and

$$x_{cO}[n] = \begin{cases} 0 & n = 0 \\ (x[n] - x[N-n])/2 & 1 \leq n \leq N-1 \end{cases}$$

In such a case, it is true that

$$DFT(x_{cE}[n]) = \text{Re}(X[k]) \tag{3.11}$$

and

$$DFT(x_{cO}[n]) = \text{Im}(X[k]) \tag{3.12}$$

Example 3.7.   Write a short script to verify Eqs. (3.11) and (3.12).

The following will suffice; different sequences can be substituted for x if desired.

x=[1:1:8]; subx = x(1,2:length(x)); xe = 0.5*(subx + fliplr(subx));
xo = 0.5*(subx - fliplr(subx)); xeven = [x(1), xe], xodd = [0, xo];
DFTevenpt = fft(xeven), ReDFTx = real(fft(x))
N=length(x); var = sum([(DFTevenpt-ReDFTx)/N].^2)

3.9   GENERAL CONSIDERATIONS AND OBSERVATIONS

The computation of each bin value $X[k]$ is performed by doing a CZL (correlation at the zeroth lag) between the signal sequence of length N and the complex correlator

$$\cos[2\pi kn/N] - j\sin[2\pi nk/N]$$

Computation of the DFT using (3.7) directly is possible for smaller values of N. For larger values of N, a **Fast Fourier Transform (FFT)** algorithm is usually employed. A subsequent section in this chapter discusses the FFT.

3.9.1   BIN VALUES
- If $x[n]$ is real-valued only, then $X[0]$ is real-valued only and represents the average or DC component of the sequence $x[n]$.

- If $x[n]$ is real-valued only, $X[N/2]$ is real-valued only, since the sine of the frequency $N/2$ is identically zero. For odd-length DFTs, there is no Bin $N/2$.

Example 3.8.   Compute the DFT of the sequence [1, 1].

Note that for a 2-point DFT, the only frequencies are DC and 1-cycle (which is the Nyquist rate, $N/2$). Both Bins 0 and 1 are therefore real. Bin 0 = **sum(([1,1]).*cos(2*pi*(0:1)*0/2))** = 2 (not scaling the DFT by $1/N$). For Bin 1, we get Bin 1 = **sum(([1,1]).*cos(2*pi*(0:1)*1/2))** = 0 (intuitively, you can see that [1,1] is DC only, and has no one-cycle content).

Example 3.9. Demonstrate the conjugate-symmetry property of the DFT of the real input sequence

$$\textbf{cos(2*pi*1.3*(0:1:3)/4) + sin(2*pi*(0.85)*(0:1:3)/4)}$$

We make the call

$$\textbf{fft([cos(2*pi*1.3*(0:1:3)/4) + sin(2*pi*(0.85)*(0:1:3)/4)])}$$

which returns the DFT as

$$\textbf{[1.611, (1.133 - 0.291i), 0.120, (1.133 + 0.291i)]}$$

We note conjugate symmetry for Bins 1 and 3, where Bin 3 is an alias of Bin -1. Note that Bins 0 and 2 (i.e., $N/2$) are unique, read-only bins within a single period of k and do not have negative frequency counterparts.

3.9.2   PERIODICITY IN N AND K

For the sake of simplicity, the following discussion assumes that N is even (the DFT of an odd-length sequence differs chiefly in that there is no Bin $N/2$, and the range of k must be adjusted accordingly).

The DFT is periodic in both n and k, meaning that if n assumes, for example, the range N to $2N - 1$, or $2N$ to $3N - 1$, etc., the result will be the same as it would with n running from 0 to $N - 1$. In a similar manner, k running from N to $2N - 1$ yields the same result as k running from 0 to $N - 1$. Or, using symmetrical k-indices, with k from $-N/2 + 1$ to $N/2$, k could instead run from $N/2 + 1$ to $3N/2$, or from $3N/2 + 1$ to $5N/2$, etc.

Stated formally, we have

$$x[n + N] = x[n]$$

and

$$X[k + N] = X[k]$$

These two relationships are true for all real integers n and k. This is the result of the periodicity of the complex exponential, $\exp(j2\pi nk/N)$ over 2π. Thus, we have

$$\exp(j[2\pi k(n + N)/N]) = \exp(j[2\pi nk/N + 2\pi k]) = \exp(j[2\pi nk/N])$$

and likewise

$$\exp(j[2\pi n(k + N)/N]) = \exp(j[2\pi nk/N + 2\pi n]) = \exp(j[2\pi nk/N])$$

- Figure 3.8 illustrates the periodicity of the DFT. The DFT of a signal of length 32 has been computed for bins -15 to 32. A complete set of DFT bins may be had by using either $k = 0$ to $N - 1$ (standard MathScript method) or $k = -N/2 + 1$ to $N/2$. By computing 48 bins, as was done here, instead of the usual 32, both arrangements may be seen in Fig. 3.8.

- In plotting a MathScript-generated DFT, a simple command (for example) such as

$$y = ones(1,32); stem(abs(fft(y)))$$

will plot the bins with indices from 1 to N. In order to plot the bins with indices from 0 to $N - 1$, write

$$x = 0:1:length(y) - 1; stem(x, abs(fft(y)))$$

- For the case of k from $-N/2 + 1$ to $N/2$, $X[k]$ and $X[-k]$ show conjugate symmetry, i.e., conjugate symmetry is shown about Bin[0]. For example, at plot (a) of Fig. 3.8, Re($X[1]$) = Re($X[-1]$), showing even symmetry, and at plot (b), Im($X[k]$) = -Im($X[-k]$), showing anti-symmetry.

- For the case of k from 0 to $N - 1$, the negative bins are aliased as bins having values of k greater than $N/2$. Thus, for the example shown in Fig. 3.8, for $k > 16$ (i.e., $N/2$), note that $X[k - N]$ = $X[k]$. For $k = 17$, the equivalent bin is Bin[17-32] = Bin[-15], Bin[18] = Bin[-14],...and finally, Bin[31] = Bin[-1]. Note that the periodicity extends to positive and negative infinity for all real integer values of k. Bin[32], for example, is equivalent to Bin[0], as is Bin[64] (not shown), etc.

- For the asymmetrical bin arrangement, conjugate symmetry is had between Bin[N/2 - m] and Bin[N/2 + m]. For example, as seen in Fig. 3.8, Bin[15] is the complex conjugate of Bin[17], which is equivalent to Bin[17-32] = Bin[-15].

Example 3.10.   Demonstrate the periodicity of k for the length-three DFT of the signal $[1, -1, 1]$.

An easy way to proceed is to make this call, letting k assume various values.

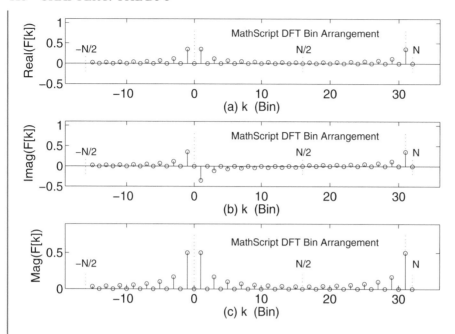

Figure 3.8: (a) The real part of the DFT of a signal, evaluated from $-N/2+1$ to N where $N = 32$; (b) The imaginary part of the DFT of the same signal; (c) The magnitude of the DFT of the signal. See the text for details of the "MathScript DFT Bin Arrangement."

$$k = 0; y = \text{sum}(\exp(-j*2*pi*(0:1:2)*k/3).*([1,-1,1]))$$

Bin 0 is, of course, real only, and there is no Bin $N/2$ since $N/2 = 1.5$. Bin 2 should prove to be the same as Bin -1. The following answers are obtained by making the calls using the values of k shown:

k	$\text{Re}(X[k])$	$\text{Im}(X[k])$
0, 3	1	0
1, 4	1	$1.7321i$
2, 5	1	$-1.7321i$
$-4, -1$	1	$-1.7321i$

3.9.3   FREQUENCY MULTIPLICATION IN TIME DOMAIN

Suppose that the time domain sequence we have been considering is multiplied by a periodic sequence with a certain frequency k_0. Let's compare the DFT of the original (unmultiplied) sequence and the DFT of the sequence after it has been multiplied (before sampling) by some frequency k_0.

Before proceeding, let's try to see what to expect. From trigonometry, we know

$$\sin(\alpha)\sin(\beta) = \frac{1}{2}\cos(\alpha - \beta) - \frac{1}{2}\cos(\alpha + \beta)$$

or

$$\cos(\alpha)\cos(\beta) = \frac{1}{2}\cos(\alpha - \beta) + \frac{1}{2}\cos(\alpha + \beta)$$

and so on for various combinations of cosine and sine. When you multiply two frequencies in the time domain, the resultant signal contains the sum and difference of the original frequencies.

With this in mind, consider Fig. 3.9, which shows at (a) a cosine of frequency five; its DFT (real only, since the time domain signal is a cosine) at (c) shows spikes at frequencies 5 and -5 (Bin 31 is equivalent to Bin -1, Bin 30 is equivalent to Bin -2, etc.).

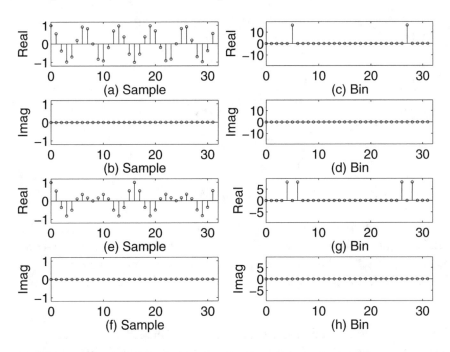

Figure 3.9: (a) and (b) Real and Imaginary parts of a complex impulse response or sequence, consisting of five cycles of a cosine; (c) and (d) Real and Imaginary parts of the DFT of the impulse response shown collectively by (a) and (b); (e) and (f) Real and Imaginary parts of the product of a five-cycle cosine with a one-cycle cosine; (g) and (h) Real and Imaginary parts of the DFT of the impulse response shown collectively by (e) and (f).

At (e), the cosine of five cycles has been multiplied by a cosine of one cycle. Thus, the expected frequencies in the new time domain signal would be 5 ± 1, and indeed, we see spikes in the DFT (plot (g)) at frequencies 4, 6, and -4 and -6 (the two latter are in their aliased positions).

In Fig. 3.10, we initially show the time domain signal as a cosine of one cycle at (a); the DFT (at (c)) shows frequencies of 1 and -1, as expected. At (e), the time domain signal is now the product of a cosine of one cycle and a cosine having five cycles. We would thus expect frequencies of 1 ± 5 = 6; −4.

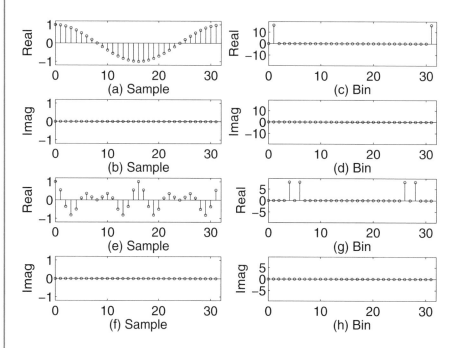

Figure 3.10: (a) and (b): Real and Imaginary parts of a complex impulse response or sequence, consisting of one cycle of a cosine; (c) and (d): Real and Imaginary parts of the DFT of the impulse response shown collectively by (a) and (b); (e) and (f): Real and Imaginary parts of the product of the one-cycle cosine with a five-cycle cosine; (g) and (h): Real and Imaginary parts of the DFT of the impulse response shown collectively by (e) and (f).

The DFT (all real) at plot (g) shows frequencies at 4, 6, 26 (= -6), and 28 (= -4), as expected.
Note that the DFT bin values in Figs. 3.9 and 3.10 were computed using MathScript, which does not scale the forward (i.e., DFT) transform by $1/N$.

3.10   COMPUTATION OF DFT VIA MATRIX

By setting up a matrix W each row of which is a DFT complex correlator, and multiplying by the signal (in column vector form), we can get the DFT.

Example 3.11.   Compute the DFT of the sequence $x[n] = [1, 2, 3, 4]$ using the matrix method and check using the MathScript function fft.

We construct the matrix as

x = [1 2 3 4]; N=length(x); nkvec = 0:1:N-1;
W = exp(nkvec'*nkvec).^(-j*2*pi/N); dft = W*x',
mfft = fft(x)

If x is complex, care must be taken as MathScript automatically conjugates a vector when it is transposed. In such case this code, which restores the signs of the imaginary parts of x after transposing, gives proper results:

x = [(1+j) 2 (3+j) 4]; N=length(x); nkvec = 0:1:N-1;
W = exp(nkvec'*nkvec).^(-j*2*pi/N); dft = W*conj(x'),
mfft = fft(x)

3.11 DFT OF COMMON SIGNALS

We'll investigate the DFT of a number of standard signals using a series of examples. All of the scripts described in the examples below scale the DFT by $1/N$.

Example 3.12. Compute and display the DFT of a square wave synthesized from a finite number of harmonics.

Figure 3.11 shows the computation window generated by the script call

LVDFTCompute(0)

as it appears for $k = 0$. The test signal at (e) is a square wave constructed with a truncated series of odd-only harmonics as

$$W = \sum_{k=1}^{N/4} (1/(2k-1)) \sin(2\pi n(2k-1)/N)$$

where n runs from 0 to $N - 1$ and N is the sequence length.

The real and imaginary components (i.e., cosine and sine) of each complex correlator are displayed for $k = 0$ up to $k = N - 1$ (press any key to compute and display the next $F[k]$). Correlations at the zeroth lag (CZLs) are done between each complex correlator and the test signal. The real and imaginary parts of $F[k]$, as well as the magnitude, are plotted as each bin is computed.

The script uses $N = 32$ and hence the test signal contains the frequencies 1, 3, 5, 7, 9, 11, 13, and 15 with amplitudes inversely proportional to frequency.

The code that computes each bin value $F[k]$ is

F(k+1) = (1/N)*sum(Signal.*exp(-j*2*pi*t*k))

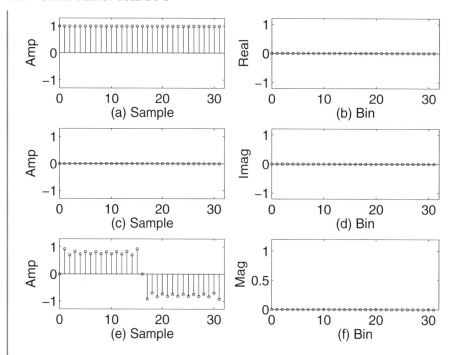

Figure 3.11: (a) and (c): Real and Imaginary correlators for Bin 0, i.e., frequency zero cosine and sine (over 32 samples), respectively; (b) and (d): Real and Imaginary parts of DFT for waveform at (a) and (c), initialized to zero with Bin 0 plotted; (e) Truncated square wave serving as test signal; (f) Magnitude of DFT, based on real and imaginary parts plotted in (b) and (d).

where $n = 0{:}1{:}N -1$ and $t = n/N$. The bins for the expression above are indexed from 1 to N as needed by MathScript, so the DC bin is F(1) rather than F(0), etc. This is corrected in the plot by plotting the output array against the vector $[0{:}1{:}N -1]$, such as by the code line

plot([0:1:N-1], F(1,1:N))

When first making the call **LVDFTCompute(0)**, the initial plot displays information for Bin 0 (i.e., $k = 0$). The real or cosine correlator is identically one for 32 samples, and the imaginary or sine correlator is identically zero for 32 samples, and the net DFT value is zero. If you are running the script, press any key, and k will be set to 1, and the result is shown in Fig. 3.12.

Note that in the complex DFT, the positive frequencies are tested with an inverted sine wave (this is the result of the negative sign in front of the j in the formula), and the negative frequencies are tested with a noninverted sine wave. Actually, it need not be this way; the negative sign in front of the j could just as well have been positive. However, for the complex DFT, the sign in front of the j used in the Inverse DFT must be opposite to the sign used in the DFT.

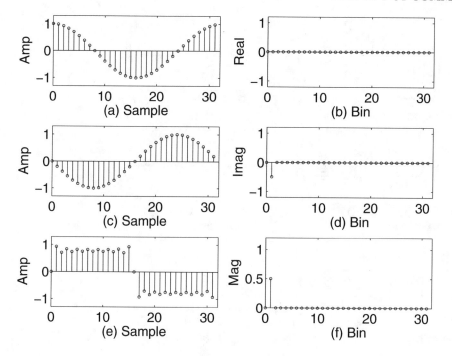

Figure 3.12: (a) and (c): Real and Imaginary correlators for Bin 1, i.e., one cycle (over 32 samples) cosine and sine, respectively; (b) and (d): Real and Imaginary parts of DFT for waveform at (a) and (c), initialized to zero with Bins 0 and 1 plotted; (e) Truncated square wave serving as test signal; (f) Magnitude of DFT, based on real and imaginary parts plotted in (b) and (d).

In Fig. 3.13, we see the final result for performing the DFT on the truncated square wave. There are a number of things to observe. First, the square wave is an odd function, and is made of a series of odd harmonics with amplitudes inversely proportional to the harmonic number. Hence, the real part of the DFT is zero.

Let's analyze the lower right plot of Fig. 3.13, which is the magnitude of the DFT.

Bin 1 (the fundamental) should have an amplitude of 1, Bin 2 should be zero (it's an even harmonic), Bin 3 should be 0.33, and so forth. Note instead that the values are half this since both positive and negative frequency bins with the same magnitude of k contribute to the reconstructed waveform when using the inverse DFT. Hence, when we add Bins 1 and 31 together, we get the required magnitude of 1 for the first harmonic ($k = 1$), 0.33 for the 3rd harmonic, 0.2 for the 5th, harmonic, etc. (later in the chapter we'll demonstrate the truth of this mathematically in a more detailed discussion of the IDFT). Recall that Bin 31 is the equivalent of Bin −1 when k runs from 0 to $N − 1$. Recall that for all Bins of the Real DFT (other than Bins 0 and $N/2$), it was necessary to double the values of the reconstructed harmonics since only positive frequencies were used in the original set of correlations. From this it can be seen that mathematically, the Real DFT is a sort of

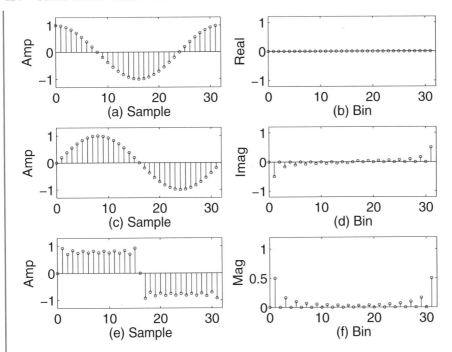

Figure 3.13: (a) and (c): Real and Imaginary correlators for Bin -1, shown after computing all bin values; (b) and (d): Real and Imaginary parts of DFT for waveform at (a) and (c), with all bin values plotted; (e) Truncated square wave serving as test signal; (f) Magnitude of DFT, based on real and imaginary parts plotted in (b) and (d).

corollary or special case of the complex DFT, which possesses a symmetry that allows the scaling value to be equal for all bins.

Example 3.13. Compute the DFT of a sawtooth wave.

The script (see exercises below)

LVxDFTComputeSawtooth

computes the DFT, step-by-step, of a truncated harmonic sawtooth which is synthesized using

$$W = \sum_{k=1}^{N/2} (1/k) \sin(2\pi nk/N)$$

where n runs from 0 to $N-1$. Note that both odd and even harmonics are included. The result from running the script through all bin values is shown in Fig. 3.14.

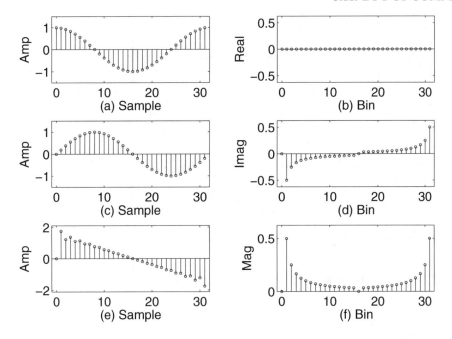

Figure 3.14: (a) Real correlator, Bin 31; (b) Real part of DFT, Bins 0-31; (c) Imaginary correlator, Bin 31; (d) Imaginary part of DFT, Bins 0-31; (e) Bandlimited sawtooth wave (input signal the DFT of which is shown in plots (b), (d), and (f)); (f) Magnitude of DFT of sawtooth shown in plot (e).

Example 3.14. Compute the DFT of a test signal using symmetric bin values.

In this case, k runs from $-N/2 + 1$ to $N/2$ rather than from 0 to $N - 1$. The result of computation is exactly the same as for asymmetrically valued k, except for the arrangement of the output. The script (see exercises below)

LVxDFTComputeSymmIndex

computes the symmetrically-indexed DFT of a bandlimited square wave, constructed from a limited number of harmonically-related sine waves.

Figure 3.15 shows the result. In the symmetric-index DFT, negative and positive frequencies per se are used and displayed in the output as such; if N is an even number, the values of k run from $-N/2+1$ to $N/2$. Other than the arrangement of bin values in the output display, the values correspond exactly to the values obtained using the nonsymmetrical system, where k runs from 0 to $(N - 1)$. Bin $(N - 1)$, for example, is identical to Bin (-1) of the symmetric-index system, Bin $(N - 2)$ is equivalent to Bin (-2), and so forth.

Example 3.15. Compute the DFT, and graph the results, for a unit impulse signal.

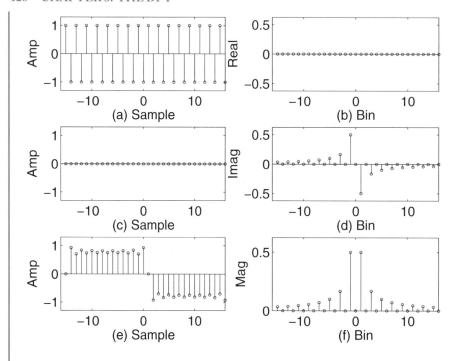

Figure 3.15: (a) and (c): Real and Imaginary correlators for Bin 16; (b) and (d) DFT with all Bin values plotted, using symmetrical Bin index; (e) Test truncated harmonic square wave; (f) Magnitude of DFT based on real and imaginary parts from (b) and (d).

In the continuous domain, the frequency spectrum of an impulse is flat, all frequencies being present with equal amplitude. In a sampled data set, we use the Unit Impulse, and all possible frequencies (positive and negative) within the sequence length are present with equal amplitudes.

The script (see exercises below)

LVxDFTComputeImpUnBal

computes and displays the DFT, one bin at a time, of a unit impulse sequence of length 32.

Figure 3.16 shows the computation completed up through bin 16. Note that the impulse is a "1" at time zero and zero thereafter, and that the imaginary correlators for all harmonics (sine waves), are all zero at time zero, thus yielding a zero-valued imaginary part for all bins. Furthermore, the real correlators for all harmonics (cosine waves) are all valued at "1" at time zero, so the DFT coefficient (bin value) for every bin is the same, and equal to $(1/32)(1)(1) = 0.0313$. Thus, the DFT over N samples of an N-sample Unit Impulse sequence consists of an equal-amplitude harmonic series of cosines.

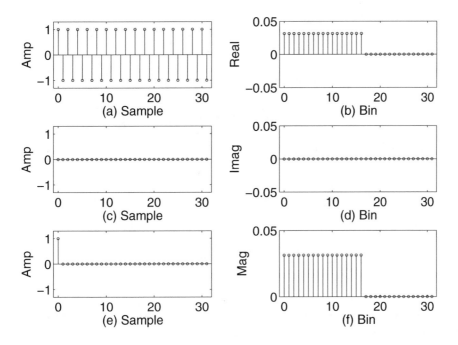

Figure 3.16: (a) and (c): Real and Imaginary correlators for Bin 16; (b) and (d): DFT with Bin values plotted up to Bin 16; (e) Unit Impulse test signal; (f) Magnitude of DFT based on real and imaginary parts from (b) and (d).

Example 3.16. Verify that, for example, a Unit Impulse sequence of length 32 is composed of a superposition of cosine waves. After synthesizing a length-32 Unit Impulse sequence, compute its DFT using the function fft and compare results to those of Fig. 3.16.

The following code synthesizes a unit impulse sequence of length SR and then computes its DFT using the function fft. The results are then plotted. Figure 3.17 shows the result. Note that the function fft does not scale the DFT coefficients by $1/N$; if the DFT employed had been one that scales the DFT coefficients, then the magnitude of the bin values in plot (b) would have been 0.0313 (1/32) rather than 1.0.

```
SR = 32; n = 0:1:SR-1; w = 0;
for ctr = 1:1:SR; w = w + (1/SR)*cos(2*pi*n/SR*(ctr-1)); end;
dftans = fft(w); figure(9); subplot(311); stem(w);
subplot(312); stem(real(dftans));
subplot(313); stem(imag(dftans))
```

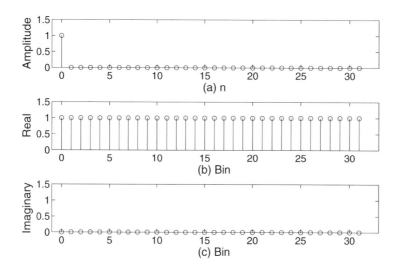

Figure 3.17: (a) A 32-point Unit Impulse sequence synthesized from cosines of length 32 and frequencies of 0 to 31; (b) Real part of DFT of same; (c) Imaginary part of DFT of same.

3.12   FREQUENCY RESOLUTION

Consider a sample sequence of length N obtained over a two second period; now correlate it with a sinusoid having one cycle over N samples. We are in effect correlating the signal sequence with a 0.5 Hz sine wave. In this particular example, the one cycle DFT correlators, sine and cosine, are equivalent to 0.5 Hz, so the next orthogonal pair (at 2 cycles) would in effect be testing the original sequence for 1 Hz, the next orthogonal pairs would be testing for 1.5 Hz, and so on.

    With this concrete example in mind, we can say that the **Frequency Resolution**, or **Bin width,** or **Bin Spacing,** available from the DFT is equal to the reciprocal of the total sampling duration T. In the example above, a total sampling duration of two seconds yielded bin center frequencies of 0 Hz, 0.5 Hz, 1 Hz, 1.5 Hz, etc. It is evident that the frequency or bin spacing, in terms of original signal frequencies, is 0.5 Hz, which is the reciprocal of the total sampling duration. We'll use the term **Bin width** in this book to mean frequency resolution or bin spacing relative to the sampled signal's actual (or original) frequencies.

    Figure 3.18 shows this concept in generic form for a total sampling duration of T seconds. We see at (a) the signal, at (b) the cosine and sine correlators for or Bin 0, at (c), the cosine and sine correlators for Bin 1, at (d) for Bin 2, and at (e), the correlators for Bin 3. These sets of correlators have frequencies over the sequence length N that go all the way up to $N/2$ for N even or $(N-1)/2$ for N odd. The equivalent signal frequency of Bin 1 is the reciprocal of T, and all higher bins represent harmonics thereof. Thus, the equivalent bin (signal) center frequencies are $0, 1/T, 2/T,$ $3/T, ... L/T$ with $L = N/2$ for N even or $(N-1)/2$ for N odd.

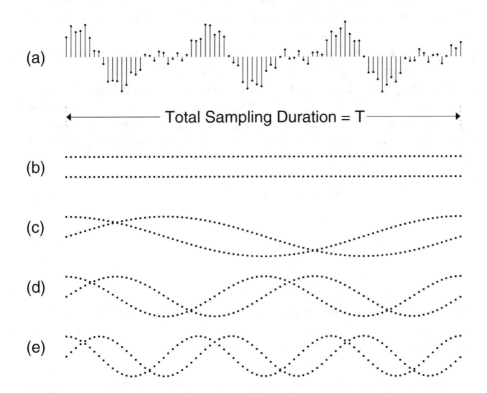

(a)

←———————— Total Sampling Duration = T ————————→

(b)

(c)

(d)

(e)

Figure 3.18: (a) Sample sequence obtained over duration T seconds; (b) Cosine and Sine correlators for frequency 0 (DC); (c) Correlators for frequency 1; (d) Correlators for frequency 2; (e) Correlators for frequency 3.

Example 3.17. A signal is sampled for 2 seconds at a sampling rate of 8 kHz, resulting in 16,000 samples. Give the equivalent signal (center) frequency of Bin 1, and give the highest bin number and its equivalent signal frequency.

The center frequency of Bin 1 in signal terms is $1/T = 0.5$ Hz. The highest bin index is $16000/2 = 8000$. The equivalent signal frequency would be $8000/T = 4000$ Hz. This is consistent with the sample rate of 8 kHz, which can only represent nonaliased frequencies up to its Nyquist rate of 4 kHz.

3.13   BIN WIDTH AND SAMPLE RATE

Note that bin width or frequency resolution is completely independent of sample rate. The maximum nonaliased passband or bandwidth is one-half the sample rate, which is independent of the total sampling duration.

For example, suppose we sampled at a rate Fs = 1024 Hz for a total of 10 seconds. With proper anti-aliasing, we can capture frequencies up to 512 Hz, and the bin width is 1/10 Hz. For a real signal, and considering zero and positive frequencies only, the DFT will have $N/2 + 1$ = 10240/2 + 1 = 5121 unique bins (the negative bins being the complex conjugate of the positive bins, and therefore not unique for real signals), spaced 0.1 Hz apart. Suppose instead we had sampled at 100 Hz for 10 seconds. In this case, the DFT would have 1000/2 + 1 = 501 unique bins, spaced 0.1 Hz apart. Thus, we have

$$Binwidth = 1/T \tag{3.13}$$

$$N = T \cdot Fs \tag{3.14}$$

$$UniqueBins = N/2 + A \tag{3.15}$$

where N is the total sampled sequence length, Fs is the sample rate or frequency, $UniqueBins$ is the total number of unique DFT bins (applies to a real signal only), T is the total sampling duration in seconds, and A is 1 if N is even and 1/2 if N is odd.

Example 3.18.   A real signal is sampled for three seconds at a sampling rate of 8000 Hz. Compute the bin width, total number of samples, and the total number of unique bins the DFT will yield.

The Nyquist limit is 8000/2 = 4000 Hz, and from Eq. (3.13), $Binwidth$ = 0.333 Hz, total samples N (from Eq. (3.14)) = (3)(8,000) = 24,000, and $UniqueBins$, from Eq. (3.15), = 24,000/2 + 1 = 12,001.

Example 3.19.   A total of 100 samples are obtained of a real signal, sampled at a constant rate for three seconds. What is the bin width, Nyquist limit, and number of unique bins?

The bin width is 0.333 Hz and the number of unique bins is 51. Since the sampling frequency is 100/3 = 33.33 Hz, the Nyquist rate is 33.33/2 = 16.66 Hz; the anti-aliasing filter should cutoff at this frequency.

Example 3.20.   A real signal is sampled for three seconds at a rate of 100 Hz. Give the bin width, the highest nonaliased frequency, the number of unique bins, and whether the following frequencies in the signal would be on-bin or off-bin in the DFT: 2.0 Hz, 4.5 Hz, 6.666 Hz.

A total of 300 samples result. The bin width is 0.333 Hz, and the bin frequencies are therefore 0, 0.333 Hz, 0.666 Hz, 1 Hz, 1.333 Hz, etc. The (nonaliased) highest frequency based on the sampling rate and proper anti-aliasing, is 50 Hz; the number of unique bins is 300/2 + 1 = 151. The frequencies 2.0 Hz and 6.666 Hz would lie squarely on-bin (in fact, on bins 6 and 20, respectively), while 4.5 Hz is off-bin.

3.14 THE FFT

There are many specialized algorithms for efficiently computing the DFT. The direct implementation of a DFT of length N requires roughly N^2 computations. The Cooley-Tukey radix-2 FFT is perhaps the most well-known of such algorithms, and requires only $N log_2(N)$ operations, which makes possible real-time computation of larger DFTs.

Example 3.21.   Assume that a certain computer takes one second to compute the DFT of a length-2^{18} sequence using an FFT algorithm. Compute how long (approximately) it would take to compute the same DFT using the direct DFT implementation.

We evaluate the ratio

$$N^2/(N \log_2(N)) = N/\log_2(N)$$

as $2^{18}/18$ = 14,564 seconds or about 4.05 hours.

3.14.1 N-PT DFT FROM TWO N/2-PT DFTS

We'll consider the simple radix-2 **Decimation-in-Time** (**DIT**) algorithm, which is based on the idea that any sequence having a length equal to a power of two can be divided into two subsequences, comprised of the even and odd indexed members of the original sequence. A DFT can then be computed for the two shorter sequences (each of which is half the length of the original sequence), and the two DFTs can be put together to result in the DFT of the original sequence. There is a computational savings achieved by doing this. You can keep dividing each subsequence into two parts until the original sequence of length N has been divided into N sequences of length one, each of which is its own DFT. From this, $N/2$ DFTs each having two bins can be assembled. The $N/2$ two-bin DFTs are then assembled into $N/4$ four-bin DFTs, and so on until a length-N DFT has been computed.

Consider the DFT ($X[k]$) of a sequence $x[n]$ of length eight samples, which can be computed in MathScript as

$$X(k) = sum(x(n).*exp(-j*2*pi*(0:1:7)*k/8))$$

The signal $x[n]$ can be rewritten as the sum of two length-4 DFTs by dividing it into its even and odd indexed samples:

$$xE = x[0:2:7]; xO = x[1:2:7]$$

which, in MathScript, would be, owing to the fact that MathScript array indices start with 1 rather than 0,

xE = x(1:2:8); xO = x(2:2:8)

For an 8-pt DFT, the complex correlator is

$$\exp(-j2\pi k(0:1:7)/8) = \exp(-j2\pi k(0:\frac{1}{8}:\frac{7}{8}))$$

For a 4-pt DFT, the complex correlator is

$$\exp(-j2\pi k(0:\frac{1}{4}:\frac{3}{4})) = \exp(-j2\pi k(0:\frac{2}{8}:\frac{6}{8}))$$

The latter expression is just the even indexed correlators (starting with index 0) for an 8-pt DFT. The odd indexed correlators for an 8-pt DFT can be obtained from the 4-pt DFT by a simple phase shift:

$$\exp(-j2\pi k(\frac{1}{8}:\frac{2}{8}:\frac{7}{8})) = \exp(-j2\pi k/8)\exp(-j2\pi k(0:\frac{2}{8}:\frac{6}{8}))$$

Then setting **n = 0:1:3**; and **Phi = exp(-j*2*pi*k/8)**, the 8-pt DFT, expressed as the sum of two 4-pt DFTs, would be

X(k) = sum(xE.*exp(-j*2*pi*(0:1:3)*k/4)) +...

Phi*sum(xO*exp(-j*2*pi*(0:1:3)*k/4))   (3.16)

At this point, we can simplify the notation by setting

$$W_N = \exp(-j2\pi/N)$$

Using this notation

$$W_N^{nk} = \exp(-j2\pi nk/N)$$

Equation (3.16) can be expressed symbolically as

$$X[k] = \sum_{n=0}^{3} x_E W_4^{(0:1:3)k} + W_8^k \sum_{n=0}^{3} x_O W_4^{(0:1:3)k}$$ (3.17)

where we have used the more standard notation x_E instead of the MathScript-suitable xE for the even subsequence, and similarly x_O for xO.

Restating Eq. (3.17) generically, we would have

$$X[k] = \sum_{n=0}^{N/2-1} x_E W_{N/2}^{nk} + W_N^k \sum_{n=0}^{N/2-1} x_O W_{N/2}^{nk}$$

which is a DFT of length N expressed as the sum of two DFTs of length $N/2$, the second of which is multiplied by a phase (or **Twiddle**) factor which adjusts the phase of its complex correlators to correspond to those of the odd-indexed values of the length-N DFT. Letting k run from 0 to $N/2$ -1, in order to account for all bins for a length N DFT, we would have

$$X[k] = X_E[k] + W_N^k X_O[k]$$

and

$$X[k + N/2] = X_E[k + N/2] + W_N^{(k+N/2)} X_O[k + N/2]$$

Note that the sequences X_E and X_O are periodic over $N/2$ samples, so $X_E[k + N/2] = X_E[k]$ and $X_O[k + N/2] = X_O[k]$ and that

$$W_N^{(k+N/2)} = W_N^k W_N^{N/2}$$

Since

$$W_N^{N/2} = \exp(-j2\pi(N/2)/N) = -1$$

we have as a result the two **Butterfly Formulas**

$$X[k] = X_E[k] + W_N^k X_O[k] \tag{3.18}$$

$$X[k + N/2] = X_E[k] - W_N^k X_O[k] \tag{3.19}$$

Figure 3.19 illustrates, in plots (a)-(c), the complex correlator and its real and imaginary parts for an 8-pt DFT, for $k = 1$. Plots (d)-(f) show the complex correlator for a 4-pt DFT, plotted as circles in all three plots. The complex values marked as stars were generated by offsetting the phase of the 4-pt DFT correlators by the phase factor $\exp(-j2\pi(1)/8)$.

3.14.2 DECIMATION-IN-TIME

This principle of breaking a length 2^n sequence into even and odd subsequences, and expressing the DFT of the longer sequence as the sum of two shorter DFTs, can be continued until only length-1 DFTs remain. A sequence of length 8, for example, is decimated into even and odd subsequences several times, in this manner.

Matrix (3.20) shows the steps for DIT of a length-8 sequence. It may be interpreted as follows: TD (1 X 8), for example, means that the sequence (whose original sample indices lie to the right),

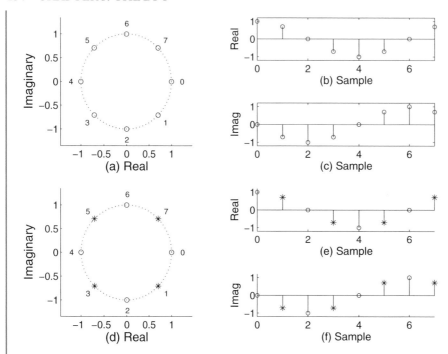

Figure 3.19: (a) The complex correlators for an 8-pt DFT, for $k = 1$; (b) Real part of the complex correlators shown in (a); (c) Imaginary part of the complex correlators shown in (a); (d) The complex correlators for an 8-pt DFT ($k = 1$), generated as the complex correlators for a 4-pt DFT (marked with circles), and the same 4-pt correlators multiplied by the phase factor $\exp(-j2\pi(1)/8)$, marked with stars.

is 1 sequence of length 8, whereas TD (2 X 4) means that the original sample indices to the right form 2 sequences of 4 samples each, and so on.

$$
\begin{array}{llllllllll}
\text{TD}(1 \text{ X } 8) & 0 & 1 & 2 & 3 & 4 & 5 & 6 & 7 \\
\text{TD}(2 \text{ X } 4) & 0 & 2 & 4 & 6 & 1 & 3 & 5 & 7 \\
\text{TD}(4 \text{ X } 2) & 0 & 4 & 2 & 6 & 1 & 5 & 3 & 7 \\
\text{TD}(8 \text{ X } 1) & 0 & 4 & 2 & 6 & 1 & 5 & 3 & 7 \\
\text{FD}(8 \text{ X } 1) & 0 & 4 & 2 & 6 & 1 & 5 & 3 & 7
\end{array}
\tag{3.20}
$$

The length-8 sequence (in the first row of the matrix) is divided into two length-4 sequences by taking its even and odd indexed parts, shown in the second row. Then the two 4-sample sequences are each split into two 2-sample sequences, forming a total of four 2-sample sequences. These are then subdivided again to form eight 1-sample sequences. The result, however, is identical to the ordering of the four 2-sample sequences. Since the last three rows are identical, there is obviously no separate step needed to generate the final two rows.

The DFT of a single sample sequence is itself, so the row TD(8 X 1) in the matrix, is also a row of eight 1-sample DFTs, labeled FD(8 X 1).

3.14.3 REASSEMBLY VIA BUTTERFLY

From there, the sequences are recombined to form four 2-pt DFTs, then two 4-pt DFTs, then one 8-pt DFT. This is done according to the Butterfly formulas given above.

In practical terms, when the DIT routine arrives at 2^{N-1} subsequences of length two, the butterfly routine can be started. For the first set of butterflies, eight 1-point DFTs are converted to four two-point DFTs using the butterfly formulas with $N = 2$ and k taking on only the single value of 0. These formulas

$$X[0] = X_E[0] + W_2^0 X_O[0]$$

$$X[0+1] = X_E[0] - W_2^0 X_O[0]$$

which simplify to

$$X[0] = X_E[0] + X_O[0]$$

$$X[0+1] = X_E[0] - X_O[0] \tag{3.21}$$

are applied in turn to the four pairs of single-point DFTs to produce the four 2-point DFTs.

At this stage we now have four 2-pt DFTs which must be assembled into two 4-pt DFTs. Using the butterfly formulas, and letting k run from 0 to 1, we have

$$X[0] = X_E[0] + W_4^0 X_O[0]$$

$$X[0+2] = X_E[0] - W_4^0 X_O[0]$$

$$X[1] = X_E[1] + W_4^1 X_O[1]$$

$$X[1+2] = X_E[1] - W_4^1 X_O[1] \tag{3.22}$$

which thus produce a new four point DFT for every four bins formed by the two bins of each of X_E and X_O.

The final step is to assemble two 4-pt DFTs into one 8-pt DFT using the butterfly formulas. A summary of the entire process in schematic form, known as a **Butterfly Diagram,** is shown in Fig. 3.20. The input to this butterfly diagram or algorithm is the time-decimated signal x (shown above the topmost row of boxes) from the final stage of time domain decimation. The values $x[0]$,

$x[4]$, etc., are reinterpreted as eight 1-pt DFTs which are labeled in pairs as $X_E[0]$ and $X_O[0]$ preparatory to combining to form four 2-pt DFTs. The applicable values for k and twiddle factors W are shown to the right of each set of butterflies.

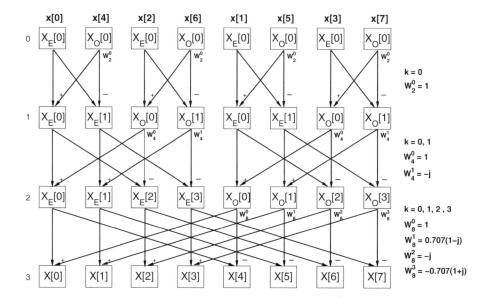

Figure 3.20: The Butterfly Diagram for a length-8 Decimation-in-Time FFT algorithm.

Example 3.22. Compute the DFT of the sequence [1 2 3 4] using the radix-2 DIT algorithm discussed above.

The first step is to decimate in time, which results in the net sequence [1,3,2,4]. Then we obtain the 2-pt DFTs of the sequences [1,3] and [2,4] which are, respectively, [4,-2] and [6,-2] by using the 2-pt butterflies (3.21). Using the 4-pt butterflies at (3.22), we combine the two 2-pt DFTs into one 4-pt DFT:

$$X[0] = 4 + 6 = 10$$

$$X[0+2] = 4 - 6 = -2$$

and

$$X[1] = -2 + (-j)(-2) = -2 + j$$

$$X[1+2] = -2 - (-j)(-2) = -2 - j$$

yielding a net DFT of

$$[10, (-2+j), -2, (-2-j)]$$

The MathScript call

$$y = \mathbf{fft}([1, 2, 3, 4])$$

yields the identical result.

3.14.4 ALGORITHM EXECUTION TIME

The DIT FFT is made more efficient by making its two components, the time decimation, and the reassembly by successive butterflies, as efficient as possible. To make the decimation more efficient, it is noted that if the samples are indexed in binary notation, the entire sequence can be properly rearranged as discussed above simply by exchanging sample values that have bit-reversed or mirror-image binary addresses. The following table shows this:

Binary	Decimal	Binary	Decimal	
000	0	000	0	
001	1	100	4	
010	2	010	2	
011	3	110	6	(3.23)
100	4	001	1	
101	5	101	5	
110	6	011	3	
111	7	111	7	

The first two columns of Table (3.23) are the forward binary count sequence and its decimal equivalent, and the third and fourth columns are the bit-reversed binary count sequence and its decimal equivalent, which can be seen as forming the proper time decimation of the 8-pt sequence. It is only necessary to exchange samples when the bit-reversed counter's value exceeds the forward counter's value. An in-place bit-reversal routine is as follows:

```
J = 0;
for I = 1:1:length(Signal)-2
k = length(Signal)/2;
while (k <= J)
J = J - k;
k = k/2;
end
```

```
J = J + k;
if I < J
 temp = Signal(J+1);
 Signal(J+1) = Signal(I+1);
 Signal(I+1) = temp;
end
end
```

The script

$$LV\_FFT(L, DecOrBitReversal)$$

graphically illustrates a test signal, its progressive decimation using either direct decimation or bit reversal, and the reassembly by butterfly. The variable *DecOrBitReversal* does the signal rearrangement by even/odd decimation if passed as 0, or by bit reversal if passed as 1. The test signal is the ramp $0:1:L-1$, which makes evaluation/visualisation of the decimation easy. Figure 3.21, for example, shows the result from the call

LV_FFT(8,0)

Since the test signal's sample values are the same as their respective (original) indices, plot (c)'s amplitude values are the same as the test sequence's original index values. Thus, we see that the sequence 0:1:7, shown at (a), has been properly decimated into the sequence [0,4,2,6,1,5,3,7], shown at (c).

A butterfly routine, compact in terms of number of lines of code, can easily be written in m-code (see exercises below) in accordance with Eqs. (3.18) and (3.19). However, a less-efficient looking, but far more efficient in practice routine to compute the butterflies from the decimated-in-time sequence is as follows:

```
Rx = real(x); Ix = imag(x);
M = log2(length(x)); LenSig = length(x);
for L = 1:1:M
 LE = 2^L; LE2 = LE/2;
 uR = 1; uI = 0;
 sR = cos(pi/LE2); sI = -sin(pi/LE2);
 for J = 1:1:LE2
 Jmin1 = J-1;
 for I = Jmin1:LE:LenSig - 1
  Ip = I + LE2;
  tR = Rx(Ip+1)*uR - Ix(Ip+1)*uI;
  tI = Rx(Ip+1)*uI + Ix(Ip+1)*uR;
  Rx(Ip+1) = Rx(I+1) - tR;
```

```
    Ix(Ip+1) = Ix(I+1) - tI;
    Rx(I+1) = Rx(I+1) + tR;
    Ix(I+1) = Ix(I+1) + tI;
   end
  tR = uR;
  uR = tR*sR - uI*sI;
  uI = tR*sI + uI*sR;
  end
 end
 x = Rx + j*Ix;
```

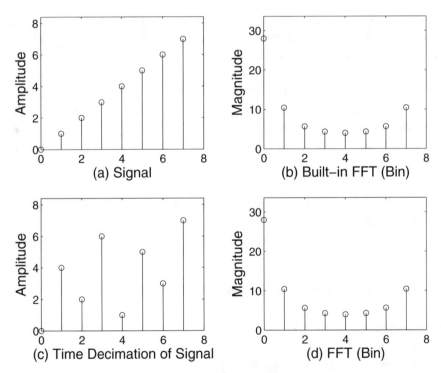

Figure 3.21: (a) Test signal, the ramp 0:1:8; (b) Built-in fft of the test signal; (c) Time decimation of the test signal; (d) FFT of test signal as performed using butterfly routines on the signal at (c).

3.14.5 OTHER ALGORITHMS

Beside the DIT algorithm, there are **Decimation-In-Frequency (DIF)** algorithms. For both types of these two basic algorithms, there are many variations as to whether or not the samples input to the butterfly routine are in natural or bit-reversed order. There are algorithms in which the input is presented in natural order and the output is in correct bin order, without any explicit bit-reversal-like reordering. There are also algorithms that do not require that the signal length be a power of two, such as relative prime factor algorithms, for example.

Reference [1] gives a very succinct explanation of the radix-2 decimation-in-time algorithm; [2] and [3] present many butterfly diagrams and discussions covering both DIT and DIF algorithms; [4] discusses factored, radix-2, radix-4, and split radix FFTs; [5] and [6] discuss matrix factorizations in addition to various standard FFT algorithms.

3.15   THE GOERTZEL ALGORITHM

3.15.1   VIA SINGLE-POLE

The simplest form of Goertzel algorithm computes a single bin value of the DFT using a single-pole IIR tuned to a DFT frequency. A finite number of iterations are performed on a given sequence of length N. To see how this works, note that each DFT bin is the result of a correlation between the signal and a complex correlator. We have also noted that correlation can be performed by convolving a signal with a time-reversed version of the correlator (i.e., correlation via convolution). If a complex correlator were time-reversed and used as an FIR impulse response, and the signal convolved with it, the output would be, when the convolution sequence reaches lag zero (perfectly overlapping the impulse response), the CZL, which in this case is the DFT value for the bin corresponding to the complex correlator.

Example 3.23.   Construct a suitable correlator for Bin 2 of a length-8 DFT to be performed on the sequence $x[n] = randn(1, 8)$, make an impulse response (in fact, a matched filter) by time-reversing the correlator, convolve $x[n]$ with the impulse response, then take the CZL as the DFT bin sought. Verify the answer using the function fft.

Code that performs the above and checks the answer is as follows:

theBin = 2; x = randn(1,8); N = length(x);
fBin = exp(-j*2*pi*(0:1:N-1)*theBin/N);
cc = fliplr(fBin); y = conv(cc,x);
CZL = y(1,N), mftx = fft(x); fftBin = mftx(theBin+1)

Now imagine that instead of performing the convolution using an FIR, it is done using an IIR having an impulse response that, over a finite number of samples, matches the complex correlator needed If, for example, the bin sought is Bin k for a length-N DFT, we set

$$p = exp(j*2*pi*k/N)$$

Let's consider the case of a length-4 sequence for which we want to compute Bin 1. If the samples of the sequence are designated $s[0]$, $s[1]$, $s[2]$, and $s[3]$, and we process the sequence with an IIR having a single pole at

$$p = \exp(j*2*pi*1/4)$$

then we can write the output values of the filter as

$s[0]$
$p^1 s[0] + s[1]$
$p^2 s[0] + p^1 s[1] + s[2]$
$p^3 s[0] + p^2 s[1] + p^1 s[2] + s[3]$
$p^4 s[0] + p^3 s[1] + p^2 s[2] + p^1 s[3]$

the last term of which, due to periodicity, is the same as

$$p^0 s[0] + p^{-1} s[1] + p^{-2} s[2] + p^{-3} s[3]$$

which is a correlation at the zeroth lag between the signal and the complex power sequence

$$\exp(-j*2*pi*1/4).\hat{} (0:1:3)$$

which is the same complex power sequence used to compute DFT Bin 1 for a length-4 DFT.

Example 3.24. Verify that, for the above example, that $p.\hat{}(4 : -1 : 1)$ is the same as $p.\hat{}(0 : -1 : -3)$.

We run the following code:

$$p = \exp(j*2*pi*1/4); \text{ans1} = p.\hat{}(4:-1:1), \text{ans2} = p.\hat{}(0:-1:-3)$$

which shows that the two expressions yield the same result.

Example 3.25. Compute Bin 3 for the 8-pt DFT of the sequence $x = [-3 : 1 : 4]$ using the Goertzel algorithm and using the function fft.

We run the code

```
Bin = 3; x = [-3:1:4]; N=length(x);
GBin = filter(1,[1 -(exp(j*2*pi*Bin/N))],[x 0]);
GoertzelBin = GBin(N+1), ft = fft(x); ftBin = ft(Bin+1)
```

You can change the values for Bin and x in the code above to experiment further.

3.15.2 USING COMPLEX CONJUGATE POLES

We can convert the single-complex-pole Goertzel IIR to an all-real coefficient IIR by using the standard technique for converting a fraction having a complex denominator to one having a real denominator. A complex FIR is created by this conversion, but its computation need only be done for the final or $(N + 1)$-th output of the IIR.

$$G(z) = \frac{(1 - p^*z^{-1})}{(1 - pz^{-1})(1 - p^*z^{-1})} = \frac{(1 - p^*z^{-1})}{1 - 2\operatorname{Re}(p)z^{-1} + |p|^2 z^{-2}}$$

Note that since p lies on the unit circle, its magnitude is 1. Note also that the real part of $\exp(j2\pi k/N)$ is $\cos(2\pi k/N)$. Thus, the simplified z-transform is

$$G(z) = \frac{1 - \exp(-j2\pi k/N)z^{-1}}{1 - 2\cos(2\pi k/N)z^{-1} + z^{-2}}$$

To implement this filter, we perform the IIR portion and obtain the N-th and $(N + 1)$-th outputs; the net output is the $(N + 1)$-th IIR output minus $\exp(-j2\pi k/N)$ times the N-th IIR output.

Example 3.26.   Compute Bin 3 for the 8-pt DFT of the sequence x = [−3 : 1 : 4] using the Goertzel algorithm as described above using a pair of complex conjugate poles.

```
Bin = 3; x = [-3:1:4]; N=length(x);
GB = filter(1,[1  -2*(cos(2*pi*Bin/N))  1],[x, 0]);
GrtzlBin = GB(N+1) - exp(-j*2*pi*Bin/N)*GB(N),
ft = fft(x); ftBin = ft(Bin+1)
```

3.15.3 MAGNITUDE ONLY OUTPUT

If only the magnitude of the DFT bin is needed, the final step of the Goertzel algorithm, which uses complex arithmetic, can be replaced with read-only arithmetic, yielding an all-real algorithm. The magnitude squared of the DFT bin as computed by the expression for *GrtzlBin* in the code above can be obtained readily as

```
MagSqGBin = GB(N+1)^2 - 2*cos(2*pi*Bin/N)*GB(N+1)*GB(N) + GB(N)^2,
AbsGBin = MagSqGBin^0.5
```

3.16 LINEAR, PERIODIC, AND CIRCULAR CONVOLUTION AND THE DFT

The mathematical formula for convolution, which we now further denote, for purposes of this discussion, as **Linear Convolution**, is, for two sequences $b[n]$ and $x[n]$ and Lag Index k, defined as

$$y[k] = \sum_{n=0}^{N-1} b[n]x[k-n] \tag{3.24}$$

Figure 3.22 shows two sequences of length eight, and their linear convolution, which has a length equal to 15 (i.e., $2N-1$).

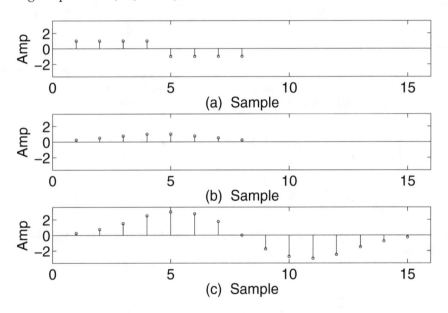

Figure 3.22: (a) Sequence 1; (b) Sequence 2; (c) Linear Convolution of Sequences 1 and 2.

3.16.1 CYCLIC/PERIODIC CONVOLUTION

Consider the case in which two sequences are both periodic over N samples. In this case, the linear convolution (when in steady-state or saturation) produces a convolution sequence that is also periodic over N samples.

Figure 3.23 shows a simple scheme wherein one sequence is just a repetition of an eight sample subsequence, and the other is a single period of an eight sample sequence. The linear convolution of these two sequences results in a periodic, or cyclic, convolution sequence when the two sequences are in saturation (steady-state response). If you look closely, you'll see that the two end (transient-response) segments of the overall convolution would actually, by themselves (removing the cyclic part of the convolution and bringing the two end segments together) constitute the linear convolution of the two eight sample sequences. Figure 3.24 shows more or less the same thing only with two periods of the second sequence. The result is that the noncyclic part of the convolution is about twice as long (nonsaturation is as long at each end as the shorter of the overall lengths of the two sequences, minus one) and the cyclic part has twice the amplitude.

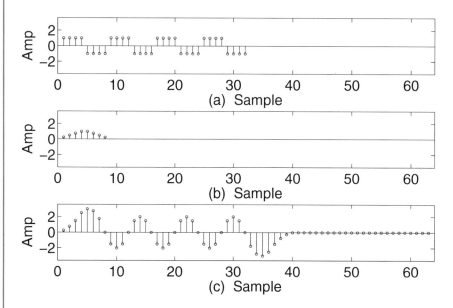

Figure 3.23: (a) A periodic sequence over eight samples; (b) An eight-sample sequence; (c) Linear convolution, showing periodicity (in saturation) of eight samples.

3.16.2 CIRCULAR CONVOLUTION

The cyclic or periodic convolution, which we have arrived at via linear convolution of periodic sequences, can also be computed (one period, that is) using a process called **Circular Convolution**, which we can define as

$$y[k] = \sum_{n=0}^{N-1} b[n](x[k-n]_N) \tag{3.25}$$

where $b[n]$ and $x[n]$ are two N-point sequences and the expression

$$x[k-n]_N$$

is evaluated modulo-N, which effectively extends the sequence $x[n]$ forward and backward so the N point convolution is a periodic convolution of $b[n]$ and $x[n]$. The term circular convolution is

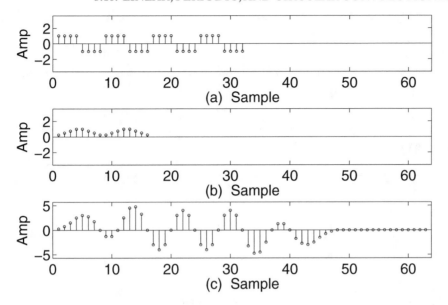

Figure 3.24: (a) A periodic sequence over eight samples; (b) Two cycles of an eight-sample sequence; (c) Linear convolution, showing periodicity (in saturation) of eight samples.

used since with the modulo index evaluation, a negative index effectively rotates to a positive one. For example, if $[k - n] = 0$, $[k - n]_N = 0$; if $[k - n] = -1$, $[k - n]_N = N - 1$; if $[k - n] = N$, $[k - n]_N = 0$, and so forth. Using the MathScript function $mod(x, N)$, for example, a Command Line call such as

$$y = mod(-2,8)$$

yields $y = 6$ which is $N - 2$. Recall that in evaluating the linear convolution formula (Eq. (3.24)), $x[k - n] = 0$ if $k - n < 0$. Note also that in evaluating $x[k - n]$ in MathScript, an offset of $+1$ is needed since 0 indices, like negative indices, are not permitted in MathScript. Thus, for example, to evaluate a circular convolution of two length-N sequences $b[n]$ and $x[n]$ in MathScript, you might employ the statement

$$y(m) = sum(b(n).*(x(mod(m - n, N) +1))); \qquad (3.26)$$

where n is the vector $1{:}1{:}N$ and m may assume values from 1 to N.

Figure 3.25 shows, at (a) and (b), two eight point sequences and, at (c), their linear convolution, having length 15 (8 + 8 - 1), and, at (d), their (8-pt) circular convolution computed using Eq. (3.25) as implemented in MathScript by Eq. (3.26). Compare this figure, plots (a), (b), and (d), to plots (a)-(c) of Fig. 3.23.

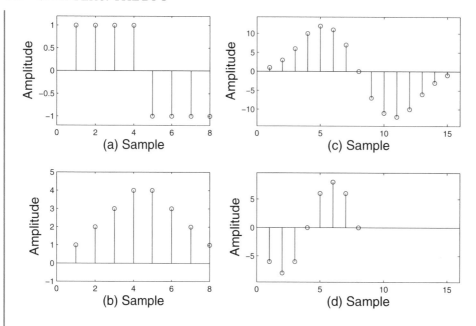

Figure 3.25: (a) First sequence; (b) Second sequence; (c) Linear convolution of sequences at (a) and (b); (d) Circular convolution of sequences at (a) and (b).

3.16.3 DFT CONVOLUTION THEOREM

If $f[n]$ and $g[n]$ are both time domain sequences that are periodic over N samples, then the DFT of the (periodic) convolution over N samples is N times the product of the DFTs of each sequence. Stated more compactly

$$DFT(f[n] \circledast_N g[n]) = N F[k]G[k]$$

where $\circledast_N$ means a periodic (or circular) convolution over N samples. Taking the inverse DFT of each side of the equation, we get

$$f[n] \circledast_N g[n] = N \cdot DFT^{-1}(F[k]G[k]) \tag{3.27}$$

which states that the periodic convolution of two sequences of length N, is N times the inverse DFT of the product of the DFT of each sequence. This assumes that the DFTs are scaled by $1/N$. Since MathScript does not scale the DFT by $1/N$, but rather scales the IDFT by $1/N$, it follows that for MathScript

$$f[n] \circledast_N g[n] = DFT^{-1}(F[k]G[k]) \tag{3.28}$$

Example 3.27.   Perform circular convolution of the sequences b = [1, 2, 2, 1] and x = [1, 0, −1, 1] using both the time domain method and the DFT Convolution Theorem, and plot the results for comparison.

The following code will suffice:

```
b = [1,2,2,1]; x = [1,0,-1,1]; N = 4; n = 1:1:N;
for m =1:1:N; CirCon(m) = sum(b(n).*(x(mod(m-n,N)+1)));
end; subplot(2,1,1), stem(CirCon)
subplot(2,1,2); y = real(ifft(fft(b).*fft(x))); stem(y)
```

3.16.4 LINEAR CONVOLUTION USING THE DFT

- A linear convolution of two sequences cannot be done using DFTs of the original sequences, but, by padding the original sequences with zeros, a problem in linear convolution can be converted to one computable as a periodic convolution, which can be computed using DFTs. For longer sequences, far greater efficiency in terms of number of computations needed can be obtained by using the DFT (implemented by FFT) technique instead of linear convolution.

Figure 3.26 illustrates how to perform a linear convolution of two length-8 sequences using circular convolution. First, pad the original sequences with zeros to a length equal to or greater than the expected length of the linear convolution. Since the ultimate goal is to use DFTs, and standard FFTs operate on sequence lengths that are powers of two, you should pick the smallest power of two that is equal to or larger than the expected linear convolution length, which is the sum of the lengths of the two sequences, less one. In this case, we pick 16 as the padded length since 15 is the required length, and 16 is the lowest power of two which equals or exceeds 15. The linear convolution of the two padded sequences is the first 15 samples of plot (c), while 16-pt circular convolutions are shown in (d) and (e), the first 15 samples of each being equivalent to the linear convolution. The circular convolution at (d) was performed using Eq. (3.25), while the circular convolution at (e) was performed using Eq. (3.28).

Example 3.28.   Consider the two sequences [1, 2, 2, 1] and [1, 0, −1, 1]. Perform the linear convolution, then obtain the same result using circular convolution. Then verify the second computation using the DFT Convolution Theorem. Plot each of the three computations on a separate axis of the same figure for comparison.

We perform and plot the linear convolution with the call

figure(120); subplot(3,1,1); stem(conv([1,2,2,1], [1,0,-1,1]))

For the circular convolution, we first pad both sequences with zeros to at least a length which is the sum of the lengths of the two (unpadded) sequences minus one. We then evaluate using

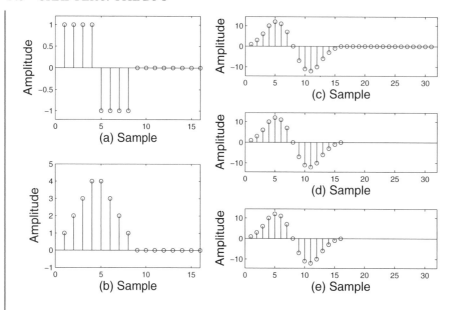

Figure 3.26: (a) First sequence, padded to length 16; (b) Second sequence, padded to length 16; (c) Linear convolution of sequences at (a) and (b); (d) Circular convolution of sequences at (a) and (b) using direct formula; (e) Circular convolution of sequences at (a) and (b) using the DFT technique.

Eq. (3.26). We define the sequences and their padded length, perform and display the computation, and verify this using the DFT Convolution Theorem with this script

```
b = [1,2,2,1,zeros(1,3)]; x = [1,0,-1,1,zeros(1,3)];
N = 7; n = 1:1:N;subplot(3,1,2)
for m =1:1:N; CirCon(m) = ...
sum(b(n).*(x(mod(m-n,N)+1))); end;
stem(CirCon); subplot(3,1,3);
y = real(ifft(fft(b,8).*fft(x,8))); stem(y(1:7))
```

Note that the DFTs have been specified as having a length equal to the smallest power of two equal to or greater than the minimum length necessary to result in a linear convolution, which is 4 + 4 -1 = 7, yielding a DFT length of eight.

3.16.5 SUMMARY OF CONVOLUTION FACTS

- Two sequences of length N when linearly convolved, yield a linear convolution of length $2N - 1$ (see Fig. 3.22).

- Two sequences, each periodic over N samples (at least one of which contains many periods) when linearly convolved, result in a convolution sequence which, in steady-state, is periodic over N samples, and is hence called a Periodic Convolution (see Fig. 3.24).

- One period of a periodic convolution can be computed using the Circular Convolution formula, Eq. (3.25).

- A linear convolution of two length-N sequences can be performed by padding each with zeros to a length at least equal to $2N - 1$, and performing circular convolution (see Fig. 3.26).

- A circular convolution in the time domain can be equivalently performed using DFTs (see Eq. (3.27) or (3.28)).

- By padding two length-N sequences with zeros up to a power of 2 which is at least equal to $2N - 1$, a linear convolution of the unpadded sequences using DFTs can be performed. This can save computation for larger sequences due to the efficiency of the FFT.

3.16.6   THE OVERLAP-ADD METHOD

Let's consider a common situation, in which we want to convolve a filter impulse response of finite length N, and a signal sequence of much greater length.

   If both sequences are of finite length, the simplest approach is to add the lengths of the two sequences and then pad that number up to the next power of two, then take the real part of the *ifft* of the product of the DFTs of the two padded sequences.

Example 3.29.   Perform linear convolution in the time domain of a 1024 sample linear chirp and the impulse response [1, 1, 1, 1], and then perform the same computation using the DFT Convolution Theorem.

   We define the signals, compute the convolution both ways, and plot the results with this call

```
SR = 1000; ts = chirp( [0:1/(SR-1):1],0,1,SR/2);
s = [1,1,1,1]; lincon = conv(s, ts);
linconbyfft = real(ifft( fft(ts, 1024).*fft(s,1024)));
figure(14); subplot(2,1,1); plot(lincon);
subplot(2,1,2); plot(linconbyfft(1:1003))
```

Note that the linear convolution has a length equal to 1000 + 4 -1 = 1003, and thus we take the leftmost 1003 samples of the convolution-via-fft as the linear convolution.

   If one of the sequences is of indefinite length (such as a sample stream from a real-time process), an approach is to size the DFTs according to the length N of the impulse response (the shorter sequence) and to break the longer sequence into shorter subsequences of length P, each of which is convolved with the impulse response, and the partial responses overlapped with the proper offset and added. The subsequence length P might also equal N, but it could be longer as well.

Either way, the impulse response and the subsequence are padded with zeros to a length at least equal to the sum of their lengths less one, and then up to the next power of two.

As an example, we'll use an impulse response of length N as one period of an N-periodic sequence, and divide the signal sequence into subsequences, each of length N.

The following procedure is known as the **Overlap and Add Method**:

- 1) Pad the impulse response with another N zeros to make a sequence of $2N$. If $2N$ is not a power of 2, pad with additional zeros until a power of 2 length is reached (call the final padded length M).

- 2) Take the first N samples of the signal and pad them to length M.

- 3) Compute the inverse DFT of the product of the DFTs of the two sequences of length M.

- 4) The result from 3) above is a sequence of length M—take the leftmost $2N - 1$ samples, which is the proper length of the linear convolution of the original two sequences of length N. These samples form the first $2N - 1$ output samples (however, only the first N samples will remain unchanged, as the next iteration will be superposed on this first output sequence starting at sample $N + 1$).

- 5) Take the second N samples of the signal sequence, pad to length M, and use with the padded impulse response to obtain the next $2N - 1$ samples of the output sequence using the DFT method.

- 6) Superpose these $2N - 1$ samples onto the current output sequence starting at sample $N + 1$.

- 7) Keep repeating the procedure for each N signal samples, and superposing the result of the Pth computation starting at output sequence sample index $(P - 1)(N) + 1$. For example, our first computation $(P = 1)$ was added into the output sequence starting at sample 1, and our second computation (a sequence of length $2N - 1$) was added in starting at sample $(2 - 1)(N)$ $+1 = N + 1$, and so forth until all signal samples have been processed.

Example 3.30.   Perform linear convolution using the DFT and the Overlap-Add Method.

Figure 3.27 shows a time just after the beginning of the DFT convolution of an eight-sample impulse response with a much longer test signal. The test signal is divided into nonoverlapping eight sample subsequences. The impulse response, and each subsequence in its turn, is padded to length-16. Then the inverse DFT of the product of the DFTs of the impulse response and subsequence is added into the cumulative convolution, starting at the beginning sample index of the particular subsequence being computed. Figure 3.27 shows the first subsequence's contribution graphed in plot (f) as a stem plot, with the second subsequence's contribution overlaid, without yet being added.

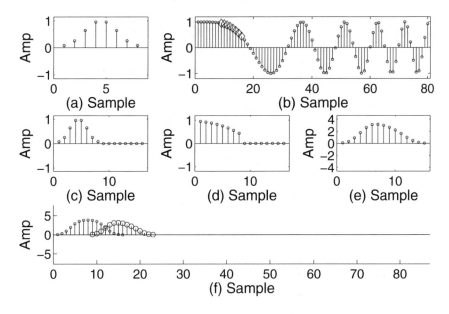

Figure 3.27: (a) An eight-sample impulse response; (b) A chirp, with second subsequence of eight-sample marked; (c) Eight-sample impulse response padded with another eight zeros; (d) Second eight-sample subsequence from (b), padded with zeros; (e) First 15 samples of circular convolution of (c) and (d), performed using DFTs; (f) Result from (e), plotted on output graph at samples 9-23, prior to being added to result from previous computation, plotted as samples 1-15.

Figure 3.28 shows the result after the second subsequence's contribution in Fig. 3.27 has been added to the composite output. Figure 3.29 shows the result after the third subsequence's contribution has been added to the composite output.

Example 3.31. Break the linear convolution of $[1, 1]$ and $[1, 2, 3, 4]$ into two convolutions, do each using the DFT, and then combine the results using the overlap-and-add method to achieve the final result.

For the first convolution, we pad $[1,1]$ and $[1,2]$ each with two zeros to form length-4 sequences, then take the real part of the inverse DFT of the product of the DFT of each, or in m-code

$$\textbf{FirstConv = real(ifft(fft([1,1,0,0]).*fft([1,2,0,0])))}$$

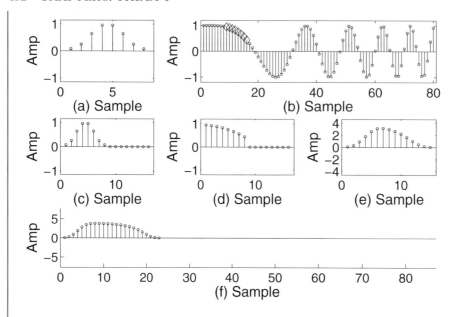

Figure 3.28: (a) Eight-sample impulse response; (b) Chirp, with second subsequence of eight samples marked; (c) Eight-sample impulse response padded with another eight zeros; (d) Second eight-sample subsequence from (b), padded with zeros; (e) First 15 samples of circular convolution of (c) and (d), performed using DFTs; (f) Result from (e), after being added to previous result, giving a valid linear convolution up to sample 16.

which yields the sequence [1,3,2,0]. The second convolution is stated in m-code as

$$\text{SecConv} = \text{real(ifft(fft([1,1,0,0]).*fft([3,4,0,0])))}$$

which results in the sequence [3,7,4,0] which is then added to the first sequence, but shifted to the right two samples:

$$
\begin{array}{cccccc}
1 & 3 & 2 & 0 & & \\
  &   & 3 & 7 & 4 & 0 \\
1 & 3 & 5 & 7 & 4 &
\end{array}
$$

This can be checked using the call

$$y = \text{conv}([1,1], [1,2,3,4])$$

which yields the identical result.

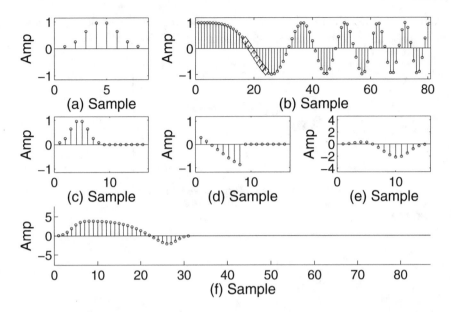

Figure 3.29: (a) Eight-sample impulse response; (b) Chirp, with third subsequence of eight-samples marked; (c) Eight-sample impulse response padded with another eight zeros; (d) Third eight-sample subsequence from (b), padded with zeros; (e) First 15 samples of circular convolution of (c) and (d), performed using DFTs; (f) Result from (e), after being added to previous result, giving a valid linear convolution up to sample 24.

3.17   DFT LEAKAGE

3.17.1   ON-BIN/OFF-BIN: DFT LEAKAGE

Consider as an example a 16 sample sequence. The test correlator frequencies that the DFT will use are -7:1:8. Any signal frequency equal to one of these will have a high correlation at the same test correlator frequency, and a zero-valued correlation at all other test correlator frequencies due to orthogonality. Such integer-valued signal frequencies are described as **On-Bin** or evoking an on-bin response, which is a response confined to a single bin.

In general, noninteger signal frequencies (i.e., not equal to any of the test frequencies 0,1,2, etc) will evoke some response in most bins. This property is usually referred to as **Leakage**, i.e., **Off-Bin** signal energy "leaks" from the closest DFT bin into other bins.

Example 3.32.   Demonstrate DFT Leakage by taking the DFT of two sequences, each of which is 64 samples long, the first of which contains a five-cycle cosine, and the second of which contains a 5.3-cycle cosine.

Figure 3.30, in plot (a), shows the DFT of the signal containing the five-cycle signal; note that only Bins ±5 are nonzero (Bin -5 appears as Bin 59 since the DFT uses $k = 0{:}1{:}N{-}1$).

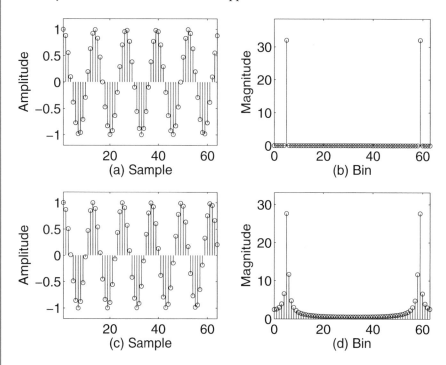

Figure 3.30: (a) Signal containing a five-cycle cosine; (b) DFT of signal in (a); (c) Signal containing a 5.3 cycle cosine; (d) DFT of signal in (c).

3.17.2   AVOIDING DFT LEAKAGE-WINDOWING

When the DFT of a sequence is taken, off-bin frequencies leak to some extent into all bins. The result can be, in the general case, a greatly diminished ability to distinguish discrete frequencies from each other when they are close. Leakage can, however, be reduced by multiplying the signal sequence by a smoothing function called a **Window** prior to taking the DFT.

Example 3.33.   Demonstrate windowing on two 64-sample sequences, one of which contains a five-cycle cosine and the other of which contains a 5.3 cycle cosine.

Figure 3.31 shows what happens when a smoothing function, in this case a Hamming window (discussed in more detail below), is used to smooth each sequence. Note that the on-bin signal is somewhat degraded, but the off-bin signal is considerably improved. In most cases, when the frequency content of a signal is unknown or random, use of a window similar to that shown in Fig. 3.31 is advantageous. This is explored in detail immediately below.

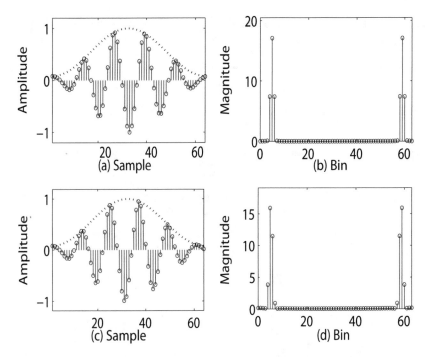

Figure 3.31: (a) Signal containing a five-cycle cosine, multiplied by a Hamming window, shown as a dashed line; (b) DFT of signal in (a); (c) Signal containing a 5.3 cycle cosine, multiplied by a Hamming window, shown as dashed line; (d) DFT of signal in (c).

3.17.3 INHERENT WINDOWING BY A RECTANGULAR WINDOW

The Rectangular window of length N, $R_N[n]$, is defined as

$$R_N[n] = \begin{cases} 1 & n = 0:1:N-1 \\ 0 & \text{otherwise} \end{cases} \tag{3.29}$$

The act of obtaining a finite sequence of length N essentially or *inherently* multiplies it by a rectangular sequence of length N, and the product of two sequences in the time domain has a frequency spectrum that is the periodic convolution of the individual frequency responses of the two sequences.

Mathematically, denoting the DTFT of $x[n]$ as $X(e^{j\omega})$, the DTFT of a given window sequence $w[n]$ as $W(e^{j\omega})$, the frequency response $X_W(e^{j\omega})$ of the windowed sequence $w[n]x[n]$ is

$$X_W(e^{j\omega}) = \frac{1}{2\pi} \int_{-\pi}^{\pi} X(e^{j\phi}) W(e^{j(\omega-\phi)}) d\phi$$

This process is shown in Fig. 3.32, in which sequences of length 2^14 were used to generate a good approximation to the DTFT, and the periodic convolution was performed numerically via circular convolution. A different look at the same signals, with a close-up of the DTFT of the window, is shown in Fig. 3.33.

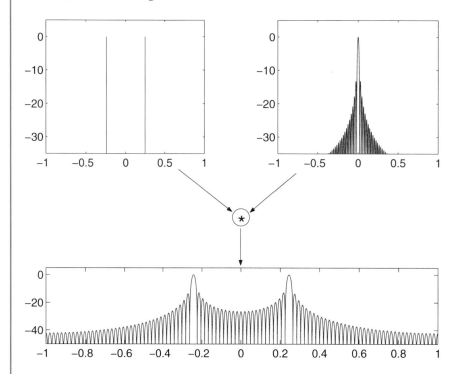

Figure 3.32: Upper left, magnitude (dB) of DTFT (via DFT of length 2ˆ14) of a cosine wave of frequency 2000; Upper right, magnitude (dB) of DTFT of Rectangular window of length 2ˆ14, with central 101 samples valued at 1.0, and all other samples valued at 0; Lower, Magnitude (dB) of the Circular Convolution of the two DTFTs shown above, giving the frequency response of the windowed cosine sequence. All horizontal axes are frequency in units of π, and all vertical axes are magnitude in dB.

We can see the effect of window length by using a rectangular window having 1001 (instead of 101) central samples valued at 1.0, the result of which is shown in Fig. 3.34. Note that with the longer window length, the frequency response of the window (plot(b)) is much narrower, and hence the net frequency response of the windowed sequence is closer to the DTFT (plot(a)) than with a shorter window. We can further emphasize this relationship with one more experiment, in which the window's central nonzero portion is shortened to 10 samples, as shown in Fig. 3.35.

As shown above, a rectangularly-windowed sequence is, due to the scalloped characteristic of its DTFT, "leaky," meaning it lets frequencies differing widely from the desired one(s) leak into the output. Fortunately, we do not have to accept this outcome. Since the net frequency response of any windowed sequence is the circular convolution in the frequency domain of the sequence's frequency response (DTFT) and the frequency response (DTFT) of the window it has been multiplied by, it

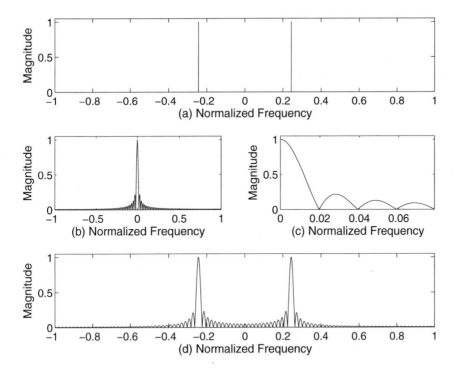

Figure 3.33: (a) DTFT magnitude of a cosine wave of length 2^14 and frequency 2000; (b) DTFT magnitude of a length-2^14 rectangular window having 101 contiguous, central values of 1.0 with all other values equal to 0; (c) A zoomed-in view of part of (b); (d) Frequency response of the windowed cosine sequence, i.e., the magnitude of the circular convolution of the DTFTs at (a) and (b).

should be possible to achieve a more desirable spectral response by choosing a window which, when convolved with the sequence's spectrum, results in less spreading out or smearing of the sequence's spectrum.

To this end, many different windows have been developed. In general, it is desirable to apply a (nonrectangular) window to a sequence prior to performing the DFT. We will explore why this is so in the following sections.

3.17.4 A FEW COMMON WINDOW TYPES

Rectangular

We defined the Rectangular window above in Eq. (3.29); Figure 3.36 shows a rectangular window having $N = 64$ which is used to window a sinusoid having 16 cycles over 64 samples; the magnitude of the DTFT is shown in plot(d). Note the heavily scalloped effect in plot(d).

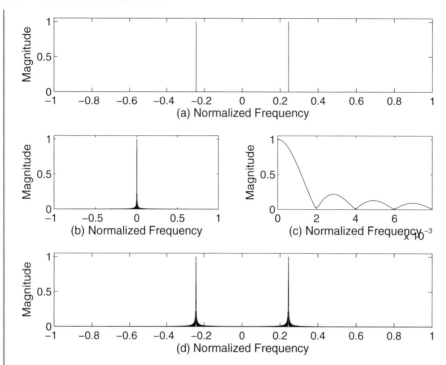

Figure 3.34: (a) DTFT magnitude of a cosine wave of length 2^14 and frequency 2000; (b) DTFT magnitude of a length-2^14 rectangular window having 1001 contiguous, central values of 1.0 with all other values equal to 0; (c) A zoomed-in view of part of (b); (d) Frequency response of the windowed cosine sequence, i.e., the magnitude of the circular convolution of the DTFTs at (a) and (b).

Hamming
The *hamming* window of length N is computed according to the formula:

$$w[n] = \begin{cases} 0.54 - 0.46(\cos[2\pi n/(N-1)]) & n = 0:1:N\text{-}1 \\ 0 & \text{otherwise} \end{cases}$$

Figure 3.37 shows the *hamming* window applied to a 16-cycle test sequence.

Blackman
The *blackman* window of length N is computed according to the following formula, in which $M = N$ -1:

$$w[n] = \begin{cases} 0.42 - 0.5\cos[2\pi n/M] + 0.08\cos[4\pi n/M] & n = 0:1:N\text{ -}1 \\ 0 & \text{otherwise} \end{cases}$$

Let's take a look at the attenuation characteristic of the *blackman* window using a log plot, which shows the fine structure of the sidelobes. From Fig. 3.38, plot(d), you can readily see that

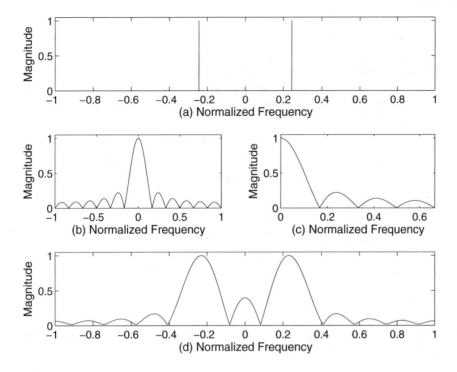

Figure 3.35: (a) DTFT magnitude of a cosine wave of length 2^14 and frequency 2000; (b) DTFT magnitude of a length-2^14 rectangular window having only 11 contiguous, central values of 1.0 with all other values equal to 0; (c) A zoomed-in view of part of (b); (d) Frequency response of the windowed cosine sequence, i.e., the magnitude of the circular convolution of the DTFTs at (a) and (b).

the scalloped sidelobes are still there, only greatly attenuated relative to the rectangular window's sidelobes. The central lobe, however, is much wider than that of either of the rectangular or hamming windows.

Kaiser
The *kaiser* window is not a single window, but a family of windows that are specified by the number of samples in the desired window and a parameter β that essentially allows you to choose the tradeoff between main lobe width and sidelobe amplitude. Figure 3.39 shows the *kaiser* window with $\beta = 5$.

You can run these scripts

LVDTFTWindowsOnly(64)

LVDTFTWindowing(64)

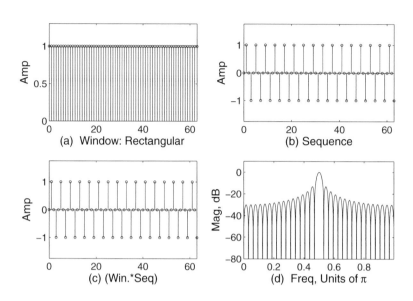

Figure 3.36: (a) 64-sample Rectangular window; (b) Impulse Response; (c) Product of waveforms at (a) and (b); (d) Magnitude (dB) of DTFT of signal at (c).

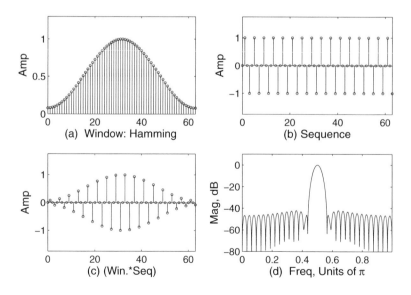

Figure 3.37: (a) 64-sample Hamming window; (b) Signal; (c) Product of waveforms at (a) and (b); (d) Magnitude (dB) of DTFT of signal at (c).

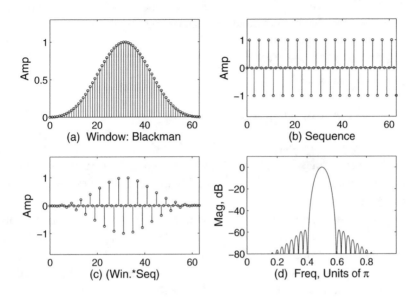

Figure 3.38: (a) 64-sample Blackman window; (b) Signal; (c) Product of waveforms at (a) and (b); (d) Magnitude (dB) of DTFT of signal at (c).

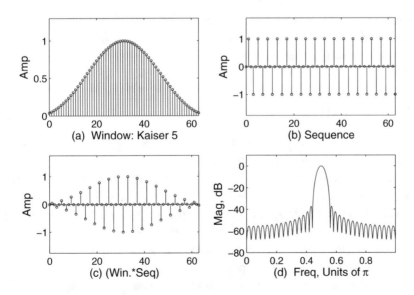

Figure 3.39: (a) 64-sample Kaiser window with $\beta = 5$; (b) Signal; (c) Product of waveforms at (a) and (b); (d) Magnitude of DTFT of signal at (c).

to view a sequence of about 15 different windows. The number in parentheses is the number of samples in the window, and you may change that number when calling the script(s).

3.17.5   DFT LEAKAGE V. WINDOW TYPE

You may run the script (see exercises below)

$$LVxWindowingDisplay(NoiseAmp, freq1, freq2)$$

which was used to generate the plots associated with this section of text. The arguments are as follows: $NoiseAmp$ is an amount of noise, $freq1$ and $freq2$ are frequencies that are immersed in the noise and which must be discriminated by performing a DFT on the composite signal (specific calls are given for the examples discussed below).

Let's do several experiments in which there are two sinusoids in white noise. In the first experiment, the sinusoids will have frequencies of 66 and 68 cycles and amplitudes of 1.0 each. These integral frequencies coincide perfectly with FFT test correlators, so such frequencies are called "**on-bin**." Frequencies that are not integers are called "**off-bin**." There is noise throughout the signal's spectrum that contributes (undesirably) to all bins, helping to blur the magnitude distinctions between bins. The script performs the same experiment for each of four different windows, namely, *rectwin*, *kaiser*, *blackman*, and *hamming*. The experiment is to construct a test waveform having the two frequencies mixed with random noise of standard deviation k, window the test waveform, and then compute the magnitude of the DFT. This is repeated 30 times, the average taken of the DFT magnitudes, and the result plotted. Then the next window is selected, the test waveform is constructed and evaluated 30 times, averaged, plotted, and so forth. In this manner, a good idea can be obtained of the average performance in noise.

Our first call will be

LVxWindowingDisplay(1,66,68)

in which the first argument is the desired noise amplitude, the second argument is the first test frequency, and the third argument is the second test frequency.

The result is shown in Fig. 3.40, plot (a). Here it can be seen that the *rectwin* window is by far the best at separating the two test frequencies, which are marked with vertical dotted lines. The *blackman* window, which has the widest central lobe, and the deepest skirt attenuation, is the poorest in this case, with the *kaiser*(5) and *hamming* windows placing between the *blackman* and the *rectwin*. Note that the two test frequencies in this case are "on-bin," and hence are orthogonal to one another and therefore cannot influence each other's DFT response.

Plot (b) of Fig. 3.40 shows the next case, close, nonintegral frequencies (66.5 and 68.6**)**. The call used to create the plot was

LVxWindowingDisplay(1,66.5,68.6)

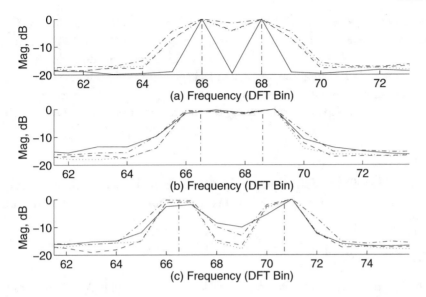

Figure 3.40: (a) Comparison of DFT response to a signal having closely-spaced on-bin frequencies with added noise for various windows–Boxcar: solid; Kaiser(5): dotted; Blackman: dash-dot; Hamming: dashed; (b) Comparison of DFT response to closely-spaced off-bin frequencies with added noise for various windows, plotted as in (a); (c) Comparison of DFT response to widely spaced, off-bin discrete frequencies with added noise, plotted as in (a). In each case the discrete test frequencies are marked with vertical lines.

In this case, it is clear that no window is adequate to separate the two frequencies; there is simply too much leakage from the off-bin test frequencies (and noise) into nearby (in fact, all) bins; the result is a "washing-out" or "de-sharpening" of the DFT response to the two frequencies.

In the third case, we extend the spacing between the off-bin test frequencies to see if the situation can be improved. The call used to create plot (c) of Fig. 3.40 was

LVxWindowingDisplay(1,66.5,70.7)

In this case, the *rectwin* is clearly the poorest performer, with *kaiser(5)* and *hamming* doing much better.

- The conventional wisdom is that the *rectwin* is the poorest performer for general purposes, due to its wideband leakage. The vast majority of frequencies in an unknown signal are likely to be off-bin (a good assumption unless the contrary is known), and hence a nonrectangular window is likely to be the better choice.

- When all frequencies being detected are orthogonal to each other, and each is "on-bin" for the DFT, and noise levels are not excessively high, the rectangular window can perform well for frequency discrimination. Each frequency, in this case, will either correlate perfectly with a DFT correlator, or not at all. Hence, the problem of leakage will exist in this case only with respect to wideband noise components in the signal which, if present, will contaminate all bins, lowering the signal-to-noise ratio and blurring the distinction among bins close to each other.

3.17.6 ADDITIONAL WINDOW USE

Windows are also commonly applied to FIR impulse responses. A later chapter (found in Volume III of the series) on basic FIR design will explore the benefits and tradeoffs associated with different windows used in FIR design.

3.18 DTFT VIA PADDED DFT

The DFT evaluates the frequency response of a sequence of length N at roughly $N/2$ frequencies such as 0, 1, etc. The DTFT can be evaluated at any number of arbitrary frequencies (such as 1.34, 2.66, 3, etc.), and thus can provide a far better estimate of the true frequency response of a sequence or system, provided a large enough number of samples are computed. By padding a sequence with zeros to a quite extended length, correlations of the sequence with many more discrete frequencies are performed, leading to a much more detailed spectrum. This is particularly useful when the padded sequence is long since, as we've seen earlier in the chapter, the FFT can efficiently compute long DFTs. The DFT/FFT method is much more efficient than the simple vector or matrix methods for computing the DTFT that were discussed earlier in the book.

Example 3.34.   Compute and display the spectrum of the impulse response $[1, \text{zeros}(1, 6), 1]$ using the DFT and a large number of samples of the DTFT.

The impulse response [1,0,0,0,0,0,0,1] has a spectrum with a number of nulls which are not adequately shown by an eight-point DFT. Computing 1024 frequency samples of the DTFT gives a more complete picture. This was achieved by padding the original impulse response with zeros to a length of 1024 samples. When using the function fft, however, you can simply specify the length of DFT to perform, and MathScript performs the zero-padding automatically. Thus, the call

$$x = [1,0,0,0,0,0,0,1]; y = \text{fft}(x,1024); \text{plot(abs(y))}$$

will compute and plot the magnitude of 1024 samples of the DTFT between normalized radian frequencies 0 to 2π.

Figure 3.41 shows the difference. The DFT values are correct for the particular frequencies they test, but the number of frequencies tested is far too small to detect the "fine" structure of the actual frequency response. The script

LV DT FT Using Padded F FT (F I RX f er F cn, F FT Length)

allows you to see the difference in spectra between a DFT having the same length as a desired impulse response (*FIRXferFcn*), and *FFTLength* samples of the DTFT. The call used to generate Fig. 3.41 was

LVDTFTUsingPaddedFFT([1,0,0,0,0,0,0,1],1024)

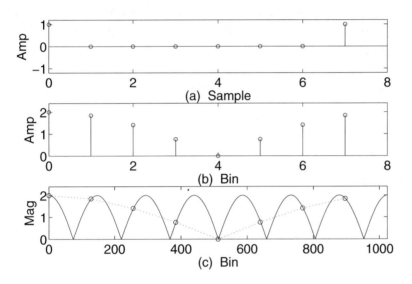

Figure 3.41: (a) Eight-sample impulse response; (b) Magnitude of eight-point DFT; (c) Magnitude of 1024-point DFT (i.e., 1024 samples of the DTFT), solid line, with eight-point DFT magnitude values plotted as stars.

Example 3.35.   An easy way to compute the frequency response at a high number of frequencies is to simply pad the input sequence with zeros and then compute the DFT of the resultant sequence.

The script

LV Padded DFT Movie(Imp Resp, Len DT FT, Comp Mode)

allows you to experiment by inputting a desired impulse response as *ImpResp*, and a desired DFT length, as *LenDTFT*. In making the script call, pass *CompMode* as *1* for automatic computation of the next sample, or *2* if you want to press any key to compute the next sample. The call used to generate Figs. 3.42 and 3.43 was

LVPaddedDFTMovie([1,0,0,0,0,0,0,1],128,1)

Figure 3.42 shows the computation for $k = 1$. The two upper plots show the sequence or impulse response under test plotted from index numbers 0 to 7, with indices 0 and 7 being valued 1 and indices 1-6 being zero (indices 8-127 are padded with zeros to compute the 128-point DFT). The cosine and sine correlators, one cycle, are also plotted, and the correlation value is plotted at index 1 in plot (c). Figure 3.43 shows the next correlation, at $k = 2$. Since the correlators are 128 samples long, k will run from 0 to 64 to cover all unique bins. The DFT of an 8-point real sequence would test five unique frequencies, i.e., $k = 0,1,2,3,$ and 4, while the 128-pt DFT gives us 65 frequencies, i.e., k = 0:1:64. Consequently, there is a much finer gradation of frequency with the longer DFT. The 65 unique bins (Bin 0, positive Bins, and Bin $N/2$) of the 128-pt DFT are shown in Fig. 3.44.

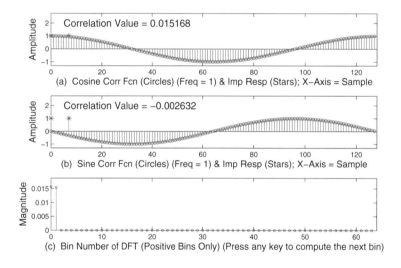

Figure 3.42: (a) Eight-sample impulse response, nonzero values plotted with stars (zero values at indices 1-6 and 8-127 not plotted), and 128-point, one-cycle cosine (real) correlator, to compute the Real part of Bin 1; (b) Eight-sample impulse response, plotted with stars (zero values at indices 1-6 and 8-127 not plotted), and 128-point, one-cycle sine (imaginary) correlator, to compute the Imaginary part of Bin 1; (c) Magnitude of DFT, plotted up to Bin 1.

Example 3.36. Estimate the harmonic content of the output of a zero-order hold DAC converting one cycle of a sine wave to an analog signal, prior to any post-conversion lowpass filtering.

Plot (a) of Fig. 3.45 shows the theoretical output of a zero-order-hold-reconstructed 1 Hz sine wave sampled at a rate of 11 Hz. Plot (b) shows a 150 sample segment of the sampled representation

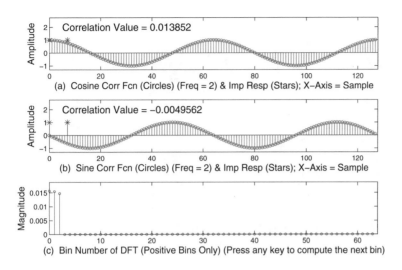

Figure 3.43: (a) Eight-sample impulse response (stars), and 128-point, two-cycle cosine (real) correlator, to compute the Real part of Bin 2; (b) Eight-sample impulse response (stars), and 128-point, two-cycle sine (imaginary) correlator, to compute the Imaginary part of Bin 2; (c) Magnitude of DFT, plotted up to Bin 2.

of the waveform at (a) (which was simulated with 1000 samples), from samples 250 to 400, which correspond to the waveform at (a) from time 0.25 second to 0.4 second (since the samples represent times 1/1000 of a second apart). A DFT is then performed on the sampled sequence, and a portion of the result (the first 60 bins) is plotted at (c). The 1000-point DFT of the sequence at (b) thus yields 1000 samples of the DTFT of the simulated 11 Hz-sampled zero-order-hold-reconstructed 1 Hz sine wave at (a). Note that the one-cycle 11-stairstep sine wave's spectrum has a high amplitude component at Bin 1, and harmonics at Bins 10 and 12, 21 and 23, 32 and 34, etc. Since the sample rate is 11 Hz, the Nyquist limit is 5.5 Hz, and this lies well below the lowest harmonics shown in plot (c) – so, with a good lowpass filter cutting off sharply at the Nyquist limit, it should be possible to eliminate all of the harmonics and have a smooth sine wave without any stairstep characteristic.

Example 3.37. Consider the sequence $[-0.1, 1, 1, -0.1]$. Figure 3.46 shows, in plots (a)-(c), the real part, imaginary part, and magnitude of the 4-pt DFT of the sequence, while plots (d)-(f) show the zero-padded 128-pt DFT of the sequence. Explain what each plot represents in terms of frequency components which make up the sequence and frequency response of the sequence.

The DFT of the four-sample sequence shows the frequency and amplitudes of the cosine and sine components used to make the sequence–i.e., the coefficients can be used to reconstruct the

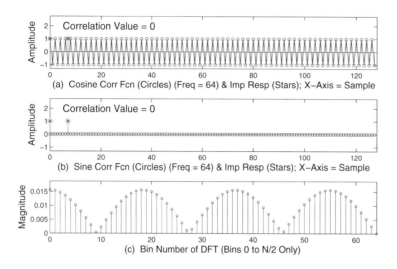

Figure 3.44: (a) Eight-sample impulse response (stars), and 128-point, 64-cycle cosine (real) correlator, to compute the Real part of Bin 64; (b) Eight-sample impulse response (stars), and 128-point, 64-cycle sine (imaginary) correlator, to compute the Imaginary part of Bin 64; (c) Magnitude of DFT, plotted up to Bin 64.

sequence exactly. Plots(d)-(f) constitute samples of the DTFT, which contain the same information of plots (a)-(c), plus much more. For example, Bin 1 of plot (a) represents one cycle over four samples, or 128/4 = 32 cycles over 128 samples. Hence, the information found for Bin 1 of plot (a) is the same as that for Bin 32 of plot (d). Using similar reasoning, Bins 0 and 2 of plot (a) would be equivalent to Bins 0 and 64 of plot (d), and analogously between plots (b) and (e), and plots (c) and (f).

Example 3.38.   Estimate the true frequency response of the cascade of three filters comprising a first IIR filter having a pole at $0.9j$, a second IIR filter having a pole at $-0.9j$, and a third (FIR) filter having the impulse response $[1, 0, 0, 0, 0, 0, 1]$ by computing 1024 samples of the DTFT of the net impulse response. Verify the answer by determining the b and a coefficients of the z-transform, using them to filter a unit impulse sequence, and then computing 1024 samples of the DTFT (using the DFT) of the unit impulse sequence. Plot the positive frequency response from both results for comparison.

The first 100 samples of the IIR impulse responses should be adequate to represent the true impulse responses since $0.9^{99} = 3*10^{-5}$, i.e., a very small number–in other words, samples of the impulse response beyond this make a negligible contribution in practical terms to the net result. The truncated impulse responses are therefore $(0.9*j).^{\wedge}(0:1:99)$ and $(-0.9*j).^{\wedge}(0:1:99)$, respectively. We

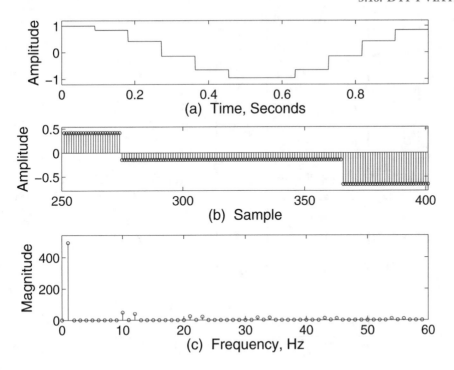

Figure 3.45: (a) A zero-order-hold-reconstructed 1 Hz sine wave prior to being lowpass filtered; (b) A 150-sample portion of the waveform at (a) simulated with 1000 samples; (c) The 1000-point DFT of the entire simulated sequence representing the waveform at (a).

can obtain the composite impulse response by convolving the two IIR impulse responses and then convolving the result with the FIR impulse response; then samples of the DTFT can be computed using a zero-padded DFT. The following code performs and plots the computation, and checks it using the *b* and *a* coefficients and the MathScript function *filter*. The results are shown in Figure 3.47.

```
y = abs(fft(conv(conv((0.9*j).^(0:1:99),(-0.9*j).^(0:1:99)),...
[1,0,0,0,0,0,1]),1024)); figure(8); subplot(211);
plot(y(1,1:513)); axis([0,513,0,inf])
a = conv([1,0.9*j],[1,-0.9*j]); b=[1,0,0,0,0,0,1]
ans = filter(b,a,[1,zeros(1,1000)]);fr=abs(fft(ans,1024));
subplot(212); plot(fr(1,1:513))
axis([0,513,0,inf])
```

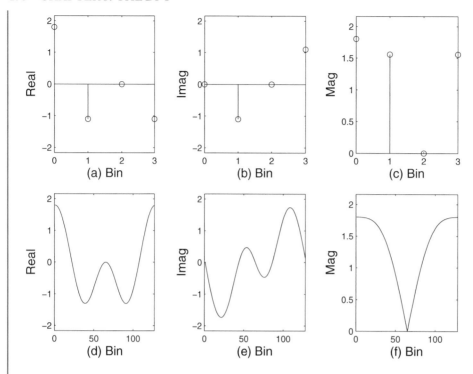

Figure 3.46: (a) Real part of 4-pt DFT; (b) Imaginary part of 4-pt DFT; (c) Magnitude of 4-pt DFT; (d) Real part of 128-pt DFT; (e) Imaginary part of 128-pt DFT; (f) Magnitude of 128-pt DFT.

3.19 THE INVERSE DFT (IDFT)

The Inverse DFT is used to convert a set of Bin values from a DFT back into a time domain sequence. If the DFT is defined as

$$X[k] = \sum_{n=0}^{N-1} x[n](\cos[2\pi kn/N] - j\sin[2\pi kn/N])$$

then the corresponding Inverse Discrete Fourier Transform (IDFT) is defined as

$$x[n] = \frac{1}{N}\sum_{k=0}^{N-1} X[k](\cos[2\pi kn/N] + j\sin[2\pi kn/N]) \tag{3.30}$$

where n runs from 0 to N - 1, as does k. Thus, the original time domain signal is reconstructed sample-by-sample.

Example 3.39. Compute the IDFT using the DFT coefficients of the sequence [2, 1, −1, 1].

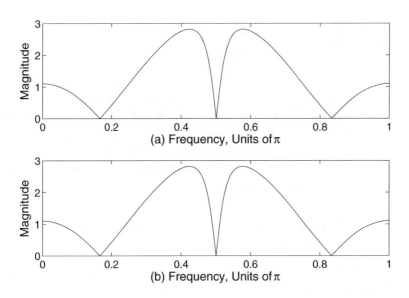

Figure 3.47: (a) Samples of the DTFT of a filter, computed as a long DFT of the convolution of two truncated IIR impulse responses and one FIR impulse response; (b) Samples of the DTFT of the same filter, estimated by computing a long DFT of the truncated impulse response of the composite filter, which was computed from the z-transform b and a coefficients.

The following code will obtain the DFT coefficients, initialize several values, then compute and display the answer:

```
s = [2,1,-1,1]; n = 0:1:3; F = fft(s); k = [0:1:3]'; Arg = 2*pi*k/4;
for n = 0:1:3; hold on;
stem(n, real(0.25*sum(F(k+1)*exp(j*Arg*n)))); end
```

Perhaps a more intuitive way of thinking about the IDFT is to generate it not point-by-point, but harmonic-by-harmonic. In this case,

$$x[0 : N - 1]_k = X[k]e^{j2\pi nk/N}$$

or using the rectangular notation,

$$x[0 : N - 1]_k = X[k](\cos[2\pi kn/N] + j \sin[2\pi nk/N])$$

where the vector $n = 0 : N - 1$, and the complete IDFT is the accumulation of all $X[k]$-weighted harmonic basis vectors $x[0 : N - 1]_k$

$$x[0:N-1] = \sum_{k=0}^{N-1} x[0:N-1]_k$$

Notice how similar the IDFT is to the DFT; they differ only by using $x[n]$ in the forward transform vice $X[k]$ in the reverse transform, a positive sign in the exponential for the reverse transform as opposed to a negative sign for the forward transform, and the scaling constant $1/N$. The negative sign may be used in the exponential in either one of the forward or reverse transform, providing the positive sign is used in the other transform, and the scaling constant may be used on either the forward or reverse transform. Furthermore, the scaling constant can be balanced as $\sqrt{1/N}$ and applied to both the forward and reverse transforms.

Example 3.40.   Devise a script to obtain the DFT of a 4-sample sequence and then reconstruct the original sequence from the DFT coefficients one harmonic at a time.

The following code computes the IDFT one harmonic at a time; the script will advance to the next computation after a one second pause:

```
s = [2,1,-1,1]; n = 0:1:3; F = fft(s); ID = zeros(1,4);
for k=1:1:4; ID = ID + 0.25*F(k)*exp(j*2*pi*(k-1)*n/4),
pause(1); end
```

3.20   COMPUTATION OF IDFT VIA MATRIX

By setting up a matrix W each column of which is an IDFT basis vector, and multiplying by the DFT (in column vector form), we can reconstruct the signal.

Example 3.41.   Compute the DFT of the sequence $x[n] = [1, 2, 3, 4]$ using the matrix method and then reconstruct $x[n]$ using the matrix method, then and check using the function ifft.

We define the sequence and the n and k vectors, obtain the DFT coefficients using MathScript, then compute the IDFT using the matrix method and check using the function $ifft$.

```
x = [(1+2*j),2,(3-2*j),4]; N = length(x); n = 0:1:N-1; k = 0:1:N-1;
MSfft = fft(x); CW = exp(n'*k).^(j*2*pi/N);
idft = (1/N)*CW*conj(MSfft'), MSifft = ifft(MSfft)
```

A symbolic rendering of this would be

$$x = \frac{1}{N}CW \cdot D \tag{3.31}$$

where **CW** is the IDFT basis vector matrix

$$\exp(n'^*k).^{\hat{}}(j*2*pi/N)$$

where $n = k = 0:1:N-1$ and **D** is the DFT vector in column form. Here, note that we have scaled by $1/N$ since the DFT has not been so scaled. Note that the conjugate must be taken of the DFT coefficients $MSiff$ when they are transposed into a column vector since MathScript automatically conjugates vectors when they are transposed.

3.21 IDFT VIA DFT

The similarity between the DFT and the IDFT can be used to redefine the IDFT so that only a single algorithm, the DFT needs to be employed to compute both the DFT and IDFT.

If the DFT is defined as

$$X[k] = \frac{1}{N} \sum_{n=0}^{N-1} x[n]e^{-j2\pi nk/N} \tag{3.32}$$

then it can be shown that the Inverse DFT is

$$x[n] = \left(\sum_{n=0}^{N-1} (X[k]^*)e^{-j2\pi nk/N} \right)^* \tag{3.33}$$

where the symbol * is used to indicate complex conjugation. If the DFT is not scaled by $1/N$ (as is true for MathScript's function fft), then Eq. (3.33) (rather than Eq. (3.32)) should be scaled by $1/N$.

Described in words, the steps to take to perform Eq. (3.33) would be to first take the complex conjugate of the DFT coefficients, then perform the DFT on them, and then take the complex conjugate of the result.

Example 3.42.   Verify the IDFT-Via-DFT concept with a four-sample complex sequence.

An easy experiment to verify Eq. (3.33) is to run the following code:

$$idft = (1/4)*conj(fft(conj(fft([1+4*j,2+3*j,3+2*j,4+j]))))$$

which returns the input complex vector (in brackets as the argument for the innermost FFT) as the value of the variable $idft$.

Example 3.43.   Compute and display the inverse DFT of a 25% duty cycle rectangle using direct implementation of the DFT to obtain the DFT coefficients, followed by computation of the IDFT using the harmonic-by-harmonic method.

Here we outline the necessary computations of the script (not provided, see description immediately following and exercises below)

LVxInvDFTComputeRect25

The IDFT portion of the script involves generating cosine and sine basis signals of various frequencies k and length N, weighted and phase shifted according to the magnitude and phase of the DFT bin coefficient for the particular value of k, in this manner:

$$x[0 : N - 1]_k = X[k](\cos[2\pi kn/N] + j\sin[2\pi kn/N])$$

where n is a vector running from 0 to N - 1. For each value of k, a complex multiplication of the DFT bin value $X[k]$ with the complex basis vector for k is performed. Thinking in polar coordinates, we are multiplying two complex vectors, each having a magnitude and a phase angle. The product is a complex exponential consisting of two waveforms, a real (cosine) and imaginary (sine). This is done for each value of k, and the results are accumulated to obtain the output.

For k = 1, the net synthesis vector would be

$$X[1](\cos[2\pi(0 : N - 1)(1)/N] + j\sin[2\pi(0 : N - 1)(1)/N])$$

If we take the real part of the above product and plot it, we would see a cosine scaled by the magnitude and shifted by the phase of $X[1]$. Similarly, taking the imaginary part of the above sequence, and plotting it, we would see a sine wave scaled and shifted by $X[1]$.

The above process is performed for k = 0 to $N - 1$, or in the case of the symmetrical DFT, $k = -N/2 + 1$ to $k = N/2$ for N even, or $-(N - 1)/2$ to $(N - 1)/2$ for N odd, and we sum all N complex synthesis vectors to get the net IDFT.

If the DFT bin values display conjugate symmetry, as they do for real input signals (which is the case here), the resultant sum of waveforms will be identically zero for the imaginary part, and the original time domain sequence will be the real part of the inverse DFT. A real input signal yields a conjugate symmetric DFT and hence a conjugate symmetric DFT returns a real sequence when reverse transformed. Of course, it also returns an imaginary sequence, but with all values equal to zero.

The script *LVxInvDFTComputeRect25* computes the DFT and then uses DFT bin values to reconstruct the original signal one harmonic (or weighted basis vector) at a time; a figure is created to display the process, computing and displaying the DFT-bin-weighted IDFT basis vectors for +k and -k simultaneously, making it easy to see that the imaginary components cancel for each bin pair, $\pm k$.

Figure 3.48, plot (a), shows a 25% duty cycle rectangular wave as the original source wave. Plots (c) and (d) show the real and imaginary parts of the DFT-bin-weighted IDFT basis vector for k = 1, plots (e) and (f) show the real and imaginary parts of the DFT-bin-weighted IDFT basis vector for k = -1, while plot (b) shows the summation of the first two weighted basis vectors (i.e., harmonics) for k = 0 and k = 1.

Figure 3.49 shows the situation up to $k = 3$. You can already see the partial IDFT starting to look like the original signal. Of course, the partial IDFT will go through a lot of appearance changes before the original signal is completely reconstructed. In Fig. 3.50, for $k = 15$, we see essentially the original signal (reconstructed) at (b), and when the last bin ($k = 16$) is reached (see Fig. 3.51), we note no additional contribution since the Bin 16 DFT coefficient was 0.

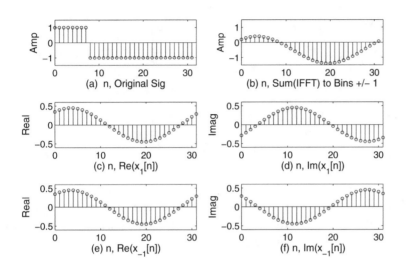

Figure 3.48: (a) Original signal; (b) Cumulative reconstructed signal, to Bin 1; (c) and (d): Real and Imaginary parts of Weighted IDFT basis vector (or harmonic) for $k = 1$; (e) and (f): Real and Imaginary parts of Weighted IDFT basis vector (or harmonic) for $k = -1$.

Example 3.44. Show mathematically how the cosine and sine components shown in plots (e) and (f) of Figs. 3.48-3.51 obtain their apparent phase shift (i.e., note that the cosine components in the figures generally do not start at zero degrees phase, nor do the sine components). Show mathematically how imaginary components cancel in the Inverse DFT if the original time domain sequence being reconstructed was real only.

Since Bins 0 and $N/2$ are real, we need only consider Bins 1:1:$(N/2 -1)$ and their negatives, i.e., Bins 1, -1, 2, -2, etc., for symmetrical indexing, or Bins 1 and $(N - 1)$, 2 and $(N - 2)$, etc., for asymmetrical indexing (i.e., when $k = 0:1:(N - 1)$). For a real input (time domain) signal, DFT Bins k and $-k$ (k not equal to 0 or $N/2$) are complex conjugates. If we let $DFT[k] = a + jb$ and $n = 0:1:N - 1$ then we can write

$$IDFT[k] = [a + jb](\cos[2\pi nk/N] + j\sin[2\pi nk/N])$$

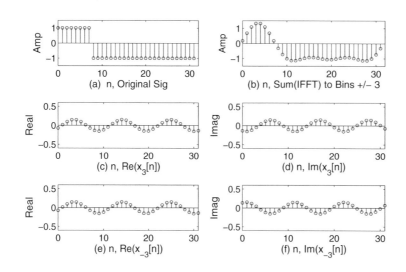

Figure 3.49: (a) Original signal; (b) Cumulative reconstructed signal, to Bin 3; (c) and (d): Real and Imaginary parts of Weighted IDFT basis vector (or harmonic) for $k = 3$; (e) and (f): Real and Imaginary parts of Weighted IDFT basis vector (or harmonic) for $k = -3$.

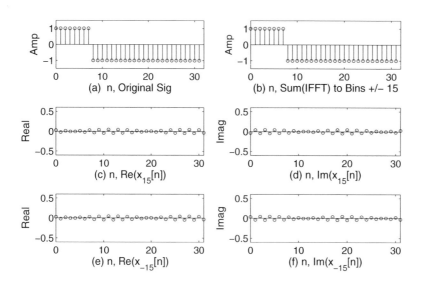

Figure 3.50: (a) Original signal; (b) Cumulative reconstructed signal, to Frequency/Bin 15; (c) and (d): Real and Imaginary parts of Weighted IDFT basis vector (or harmonic) for $k = 15$; (e) and (f): Real and Imaginary parts of Weighted IDFT basis vector (or harmonic) for $k = -15$.

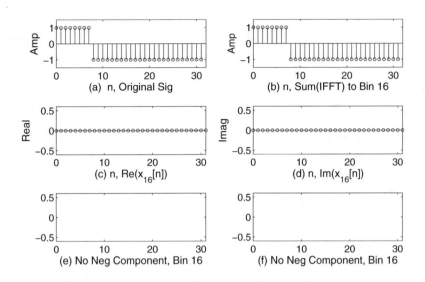

Figure 3.51: (a) Original signal; (b) Cumulative reconstructed signal, to Bin 16; (c) and (d): Real and Imaginary parts of Weighted IDFT basis vector (or harmonic) for $k = 16$; (e) and (f): (No plot as there is no Bin -16).

and

$$IDFT[-k] = [a - jb](\cos[2\pi nk/N] - j\sin[2\pi nk/N])$$

where we note that $\cos(-\theta) = \cos(\theta)$ and $\sin(-\theta) = -\sin(\theta)$. Doing the complex multiplications and adding the two resultant harmonic components, we get

$$IDFT[k] + IDFT[-k] = 2a\cos[2\pi nk/N] - 2b\sin[2\pi nk/N] \qquad (3.34)$$

We see that, for a real time domain signal, the sum of all complex DFT-Bin-weighted IDFT basis vectors (harmonics) is real. The contribution from Bins 0 and $N/2$ must also be real since the DFT Bin values themselves are real as are the IDFT basis vectors for $k = 0$ and $N/2$. It can also be seen that the reconstructed sinusoid for Bin k will have a phase determined by the values of a and b, i.e., $DFT[k]$.

A complete IDFT formula for N even using Eq. (3.34) would be

$$x[n] = Bin[0] + Bin[N/2](-1)^n + \tag{3.35}$$

$$2 \sum_{k=1}^{N/2-1} (\text{Re}(Bin[k]) \cos[2\pi nk/N] - \tag{3.36}$$

$$\text{Im}(Bin[k]) \sin[2\pi nk/N]) \tag{3.37}$$

where $n = 0{:}1{:}(N-1)$, and for N odd,

$$x[n] = Bin[0] + 2 \sum_{k=1}^{(N-1)/2} (\text{Re}(Bin[k]) \cos[2\pi nk/N] - \text{Im}(Bin[k]) \sin[2\pi nk/N]) \tag{3.38}$$

where $n = 0{:}1{:}(N-1)$.

Note initially that Eqs. (3.35) and (3.38) are only valid when the sequence $x[n]$ is real since their derivation was predicated on that premise. Notice also that the complex DFT bins have been decomposed into their real and imaginary parts, which are separately applied to the real and imaginary IDFT basis vectors (i.e., cosine and sine vectors). Note also that Bins 0 and $N/2$ get a weight of 1 and Bins 1 to $N/2-1$ (or $(N-1)/2$ if N is odd) get a weight of 2–this is essentially the real DFT/IDFT, with the real DFT coefficients being the real part of the complex DFT coefficients, etc. Note that Eqs. (3.35) and (3.38) will need to be scaled by $1/N$ if the DFT coefficients were not so scaled. Note that the imaginary component in Eqs. (3.35) and (3.38) is negative; this is because (as mentioned earlier in the book) the real DFT and real IDFT have the same signs for the imaginary component. Since in this case we started with DFT coefficients computed using the complex DFT, which uses negative imaginary correlators (i.e., exp(-j$2\pi nk/N$)), we must have a negative sign for the imaginary reconstruction components.

The script (see exercises below)

$$LVxIDFTviaPosK(TestSig)$$

uses Eqs. (3.35) and (3.38) to reconstruct a real test signal $TestSig$ from its DFT coefficients. It also reconstructs $TestSig$ using the standard function $ifft$ for comparison.

The call

LVxIDFTviaPosK([randn(1,19)])

results in Fig. 3.52.

3.22   IDFT PHASE DESCRAMBLING

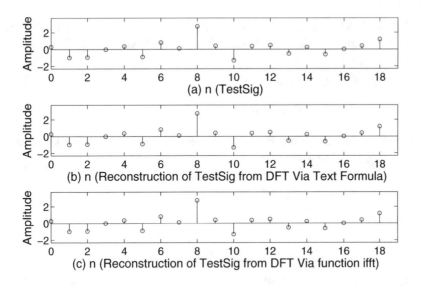

Figure 3.52: (a) Test signal; (b) Reconstruction of test signal from its DFT coefficients using Eqs. (3.35) and (3.38); (c) Reconstruction of test signal from its DFT coefficients using the standard function $ifft$.

3.22.1 PHASE ZEROING

An interesting and instructive use of the DFT/IDFT pair is to realign the phases of all the frequencies in a signal. Suppose we had a periodic signal with the phases randomly arranged, and we wanted them all to be aligned.

The phase angle of any bin can be set to zero by multiplying the bin value by its own complex conjugate, which yields a real number, and then taking the square root of the resultant real number to correct the magnitude to the original value.

Any bin may be represented as

$$a + jb = M \angle \theta$$

Multiplying by the complex conjugate, we get

$$(M \angle \theta)(M \angle (-\theta)) = M^2$$

Then to return to the correct magnitude, we take the square root. Thus, the entire operation (per bin) is

$$NewBinVal = \sqrt{(a + jb)(a - jb)}$$

3.22.2 PHASE SHIFTING

To adjust the phase of any bin to be whatever you want, simply multiply the positive frequency bins by whatever phase angle you want to shift by, and multiply the corresponding negative frequency bins by the complex conjugate of that phase angle.

Recall that a phase angle in degrees is specified as a complex number, such as

$$e^{j2\pi(\theta/360)}$$

Let's say we wanted to shift every bin by $\pi/4$ radians (45 degrees); this can be done by multiplying each positive frequency bin value by

$$e^{j2\pi(45/360)} = e^{j\pi/4} = 0.707 + j0.707$$

and each negative bin value by

$$e^{j2\pi(45/360)} = e^{-j\pi/4} = 0.707 - j0.707$$

For bin zero, it is not necessary to multiply by a phase angle, since DC has no phase. This is automatically taken care of in the script *LVxPhaseShiftViaDFT*, which we'll discuss below, by specifying the phase factor as dependent on frequency k; hence for $k = 0$ the phase factor turns out to be 1.

A script designed to demonstrate this (see exercises below) is

LVxPhaseShiftViaDFT

which generates a waveform having the harmonic amplitudes of a square wave, but totally random phases for all harmonics present. All phases are returned to zero, which produces a cusped waveform; to correct this, the phases of all bins are then shifted 90 degrees using the same technique, which places all harmonics in just the proper phase so the waveform appears as a square wave. The script *LVxPhaseShiftViaDFT* then presents a general phase-shifting demo, in which the initial derandomized result is incrementally shifted in phase until its phase has been shifted 90 degrees from its initial value of zero degrees for all frequencies. Each time the derandomized waveform is phase-shifted, the Hilbert Transform of the resultant waveform is taken and also plotted. The result is that initially, the derandomized waveform is the cusped waveform, and its Hilbert Transform is a square wave. After phase shifting by 90 degrees, the cusped waveform becomes a square wave, while its Hilbert transform becomes a cusped waveform.

Figure 3.53 shows the initial random-phase test signal, the derandomized result (in the time domain), and then the Hilbert Transform of that.

3.22.3 EQUALIZATION USING THE DFT

An interesting use of the DFT/IDFT is to analyze the impulse response of a system that distorts signals and to generate an inverse or deconvolution filter to reverse the distortion induced by the

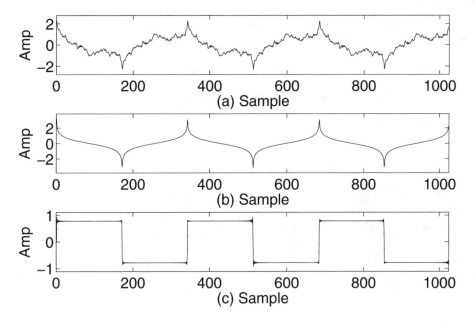

Figure 3.53: (a) Test waveform, sum of odd harmonic cosines with amplitudes inversely proportional to harmonic number and random phase; (b) Signal from (a) with all harmonics having had their phase angles set to zero; (c) Signal from (b), with all phase angles shifted 90 degrees.

system. The DFT can be used to determine the magnitude and phase characteristics of the system by taking the DFT of the system's impulse response. The reciprocal of the DFT coefficients can be used to generate a time domain filter suitable for circular convolution, or the reciprocal DFT coefficients can be multiplied by the DFT of a block of distorted signal samples, and the IDFT computed to produce the equalized (or deconvolved) signal.

The script (see exercises below)

$$LVxDFT\,Equalization(tstSig, p, k, SR, xplotlim)$$

creates a test signal of length 1024 samples, which is selected by input argument $tstSig$, and is one of several square wave trains or a chirp. A test digital impulse response to model the distorting analog channel or system is generated by multiplying random noise of standard deviation k by a decaying impulse response generated by a single real-pole IIR having a z-transform $H(z) = 1/(1 - pz^{-1})$. A DFT of length SR, which should be two or more times the length of the test sequence, is computed and its reciprocal is used to deconvolve or equalize the distorted test signal (which has been distorted by filtering the test signal using the generated test impulse response) two different ways. In the first method, the reciprocal DFT coefficients are multiplied by the DFT of the distorted signal, and the real part of the IDFT of this product produces the deconvolved signal. In the second method, a time domain impulse response is produced by taking the IDFT of the reciprocal DFT coefficients, and then circularly convolving the distorted (time domain) test signal with it.

The call

LVxDFTEqualization(3,0.9,0.2,2048,450)

results in (for example) Fig. 3.54.

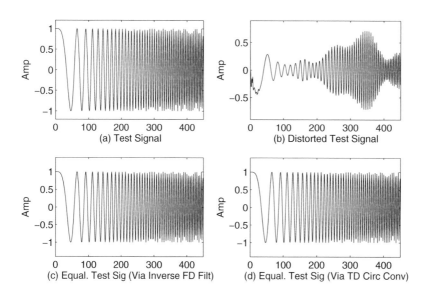

Figure 3.54: (a) Test Signal; (b) Distorted Test Signal, after being convolved with simulated signal processing channel's impulse response; (c) Deconvolved Test Signal, produced using frequency domain filtering following by the IDFT; (d) Deconvolved Test Signal, produced by using the IDFT on the reciprocal of the DFT of the channel impulse response to produce a time domain filter which is circularly convolved with the distorted test signal.

3.23   REFERENCES

[1] William L. Briggs and Van Emden Henson, *The DFT*, SIAM, Philadelphia, 1995.

[2] Richard G. Lyons, *Understanding Digital Signal Processing*, Addison Wesley Longman, Inc., Reading, Massachusetts, 1997.

[3] Alan V. Oppenheim and Ronald W. Schaefer, *Discrete-Time Signal Processing*, Prentice-Hall, Englewood Cliffs, New Jersey, 1989.

[4] John G. Proakis and Dimitris G. Manolakis, *Digital Signal Processing, Principles, Algorithms, and Applications, Third Edition*, Prentice Hall, Upper Saddle River, New Jersey, 1996.

[5] E. Oran Brigham, *The Fast Fourier Transform and Its Applications*, Prentice-Hall, Inc., Upper Saddle River, New Jersey, 1988.

[6] Charles Van Loan, *Computational Frameworks for the Fast Fourier Transform*, SIAM, Philadelphia, 1992.

[7] Steven W. Smith, *The Scientist and Engineer's Guide to Digital Signal Processing*, California Technical Publishing, San Diego, 1997.

3.24 EXERCISES

1. For the following signals, state how many total bins will be generated by the DFT, what the bin width (bin spacing) is, and how many unique bins there are. The term random is used to mean that the sequences possess no particular symmetry such as being even, odd, etc.

   (a) A real random sequence $x[n]$, obtained over a period of 2.5 seconds at a sample rate of 750 Hz.

   (b) A real random sequence comprising 101 samples obtained at a sample rate of 4000 Hz.

   (c) A complex random sequence of 499 samples obtained at a sample rate of 11025 Hz.

2. Show, using Eq. (3.7), that a) the DFT of an arbitrary real sequence $x[n]$ multiplied by a complex exponential will be shifted in frequency, and b) the DFT of an arbitrary real sequence $x[n]$ multiplied by a cosine will consist of frequency components that are the sum and difference of the original components and the frequency of the cosine.

3. (a) For a length-8 DFT, specify eight different sinusoids the DFTs of which will be real and have no leakage, i.e., the energy will be concentrated in one or two (considering both positive and negative frequencies) bins only.

   (b) For a length-7 DFT, specify seven different sinusoids the DFTs of which will be imaginary-only and have no leakage, i.e., the energy will be concentrated in one or two bins only.

4. Write a script that computes the DFT of an arbitrary real input sequence $x[n]$ of length N using the definition of the DFT (either of Eqs. (3.6) or (3.7)) but which uses symmetric indexing, i.e., the range for both n and k is $-N/2+1:N/2$ for N even and $-(N-1)/2:(N-1)/2$ for N odd. On a single figure, plot $x[n]$ and the real and imaginary parts of the DFT.

   Test the script with the input signals given below, and, verify results by creating a separate figure, with the same plots, but populate them by using the MathScript function *fft* on the various input signals. Use the function *fftshift* to bring the results into asymmetric format. For each grouping of like test signals, compare results. For the first two groupings ((a),(b) and (c),(d)), how does the length of the test sequence affect the outcome? For the last grouping, how does the number of cycles of the sawtooth affect the DFTs fundamental and harmonic series?

   (a) x = [1 zeros(1,11)]; (b) x = [1 zeros(1,51)]

(c) x = [ones(1,17)]; (d) x = [ones(1,73)]
(e) x = [-31:1:32]
(f) x = mod([-31:1:32],32); (g) x = mod([-31:1:32],16)
(h) x = mod([-31:1:32],8); (i) x = mod([-31:1:32],4)

5. Write a script that will verify the DFT shift property for an arbitrary input signal $x[n]$, i.e., verify that, for an arbitrary shift of m samples,

$$DFT(x[n - m]_N) = X(k)e^{-j2\pi mk/N}$$

More specifically, the script should receive $x[n]$ and m and 1) compute the fft of $x[n - m]_N$, 2) compute the fft of $x[n]$ and multiply it by the appropriate phase factor, and 3) plot the magnitude and phase of the results from 1) and 2) above.

6. Compute and display the magnitude of the DTFT and DFS on the same plot for the sequences below.

(a) [1,0,0,0,0,1]
(b) fir1(10,0.5)
(c) [ones(1,8)]
(d) [real(j.^(0:1:10))].*blackman(11)'

7. For the periodic sequence $x[n]$ of infinite length, one period $\tilde{x}[n]$ of which is [1,2,3,4], compute the DFS coefficients, then use the inverse DFS formula to compute $x[n]$ over the range -21 < n < 21. Plot $\tilde{x}[n]$ and the 41 reconstructed samples of $x[n]$, with proper indices for both, on the same axis.

8. For the three scripts described in the text, namely

$$LVxDFTComputeSawtooth$$

$$LVxDFTComputeSymmIndex$$

$$LVxDFTComputeImpUnBal$$

create either a single script or a single VI that will compute the DFT of any of the following signals, all of which have user-designatable length and, in cases 1) and 2), user-designatable fundamental frequency: 1) a bandlimited sawtooth; 2) a bandlimited square wave, 3) a bandlimited triangle wave, 4) a unit step. The format of DFT display should also be user-designatable as either symmetrical or asymmetrical.

The DFTs should be computable step-by-step and displayed similarly to any of Figs. 3.14, 3.15, or 3.16.

9. If a certain 4-pt sequence is represented symbolically as [a,b,c,d], derive an expression for the DFT using decimation-in-time and the 4-pt butterfly.

10. Using paper and pencil, and subsequence lengths of two samples, compute the convolution of the sequences [-1, 7] and [2, -1, 3, 0, 2, -3, 2, 4] using the overlap and add method. Proceed in the following manner: for each pair of length-2 subsequences to be convolved, pad to length-4, obtain the 4-pt DFTs of each subsequence using the expression derived in the previous problem. Compute the product of the two 4-pt DFTs and compute the inverse DFT of the product by using the DFT-by-IDFT method and the 4-pt DFT expression derived in the previous problem. Take the resulting time domain answer and add it to the cumulative result with the proper offset, etc. Check your answer by performing the time domain linear convolution using the *conv* function.

11. If a certain computer can perform a direct-implementation DFT of length 4096 in 36 seconds, approximately how long should the same computer need to perform a DFT of length 262,144 samples using a decimation-in-time FFT?

12. Write a script that will receive a sequence and perform a radix-2 decimation-in-time FFT on the sequence. Sequences that are not a power of two in length should be automatically padded out with zeros to a power-of-two length by your script. Write the script to allow the use of two methods to perform the decimation in time: 1) a direct implementation that divides vectors into two vectors, each of half the original length, and so forth as outlined in the text, and 2) the bit reversal code given in the text.

Compare the results and execution time for the sequences below. To compare results, have the script perform the decimation, and measure the execution time, for both methods. To determine whether or not both methods arrive at the same answer, take the sum of the absolute value of the differences between the two answers, divided by the sequence length. The result should be zero or, due to roundoff error, in the vicinity of 10^{-15} or less.

    (a) **Signal = ones(1,32);**
    (b) **Signal = ones(1,256);**
    (c) **Signal = ones(1,2^10);**

13. Write the code to implement the script

    function LVx_FFTDecnTimeAltCode(x,CmprMode,..
    BitRCd,BtrFCd)
    % x is the test signal; if not a power of 2 in length,
    % it will be padded with zeros so that it is. Pass
    % CmprMode as 0 to compare the time of execution
    % for various user-written m-coded FFT's to the built-in
    % fft; Pass CmprMode as 1 to compare the time of execution
    % of a direct DFT to the built-in fft. When CmprMode is
    % passed as 0, pass BitRCd as 0 to use a standard
    % bit-reversal routine (from text) or as 1 to use direct
    % decimation (user-written). Pass BtrFCd as 0 to run the
    % butterfly routine from the text or as 1 to run a user-written
    % butterfly routine. When CmprMode is passed as 1, you may

% **pass BitRCd and BtrFCd as the empty matrix [].**
% **Test calls:**
% **LVx_FFTDecnTimeAltCode([0:1:31],1,[],[])**
% **LVx_FFTDecnTimeAltCode([0:1:1023],1,[],[])**
% **LVx_FFTDecnTimeAltCode([0:1:31],0,0,0)**
% **LVx_FFTDecnTimeAltCode([0:1:1023],0,0,0)**
% **LVx_FFTDecnTimeAltCode([0:1:31],0,1,0)**
% **LVx_FFTDecnTimeAltCode([0:1:1023],0,1,0)**
% **LVx_FFTDecnTimeAltCode([0:1:31],0,0,1)**
% **LVx_FFTDecnTimeAltCode([0:1:1023],0,0,1)**
% **LVx_FFTDecnTimeAltCode([0:1:31],0,1,1)**
% **LVx_FFTDecnTimeAltCode([0:1:1023],0,1,1)**

The script described above has two general comparison modes. The first mode, specified by input argument *CmprMode* = 1, compares the execution time and numerical result from the built-in FFT (function fft) to the results from a direct DFT implementation written in m-code by yourself.

The second comparison mode, specified by input argument *CmprMode* = 1, compares execution time and numerical result from the built-in FFT to results from an FFT routine written by you, in which you specify by the input arguments how the decimation-in-time is performed, and how the butterfly routine is performed.

The two decimation-in-time routines, selected by input argument *BitRCd*, consist of the efficient bit-reversal code given in the text, and m-code written by yourself to repeatedly subdivide the sequences into even and odd parts, as described in the text. You can probably reuse the code written for the previous exercise to implement this portion of the current project.

The two butterfly routines, selected by *BtrFCd*, consist of the efficient butterfly code given in the text, and m-code written by yourself that works in accordance with Eqs. (3.18) and (3.19).

Test the completed script by performing the FFT on the sequences below, and compare (using the sum of differences method mentioned in the previous exercise) the result to that obtained using the function fft. Use the tic and toc functions in the script to determine the execution time for the built-in FFT, the direct DFT, the two decimation-in-time routines, and the two butterfly routines. For each sequence given below, vary the input arguments as necessary to check the execution time and consistency of numerical results. There are a total of five different combinations possible. In most cases, the built-in FFT will far exceed the performance of any of the m-coded FFTs, and the m-coded FFTs will far exceed the direct DFT in performance.

(a) **Signal = [0:1:31];**
(b) **Signal = ones(1,128);**
(c) **Signal = ones(1,2^8);**

14. Write a script that implements the functions of the script *LVxCyclicVLinearConv*, described below, which performs convolution using the DFT Convolution theorem. The script also performs

the convolution using the function *conv*, and compares the results for the test calls given below. Figure 3.55 shows the result from one possible script and corresponding set of displays for the call

LVxCyclicVLinearConv(39,571)

function LVxCyclicVLinearConv(ImpRespN,TestSigN)
% ImpRespN sets the length of an impulse response which
% is a cosine at normalized frequency 0.5, which is windowed with
% a Hamming window to form a smooth lowpass impulse
% response. TestSigN is the length a of chirp which is
% convolved with the lowpass impulse response using the
% DFT with padded sequences to effect linear convolution.
% Test calls:
% LVxCyclicVLinearConv(28,280)
% LVxCyclicVLinearConv(32,1024)
% LVxCyclicVLinearConv(32,1024)
% LVxCyclicVLinearConv(39,571)

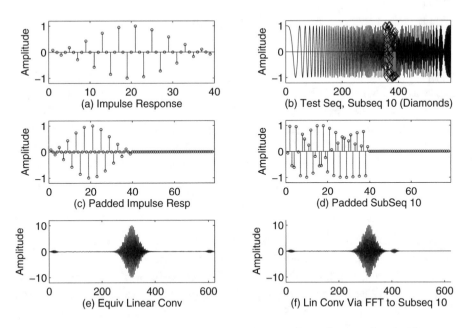

Figure 3.55: (a) Windowed impulse response; (b) Test chirp; (c) Zero-padded impulse response; (d) Subsequence 10, padded with zeros; (e) Complete linear convolution of test chirp and impulse response, performed using the function conv; (f) Partially complete convolution (through subsequence 10) of the convolution as performed using the DFT Convolution Theorem.

15. The DTMF (Dual Tone, Multiple Frequency) system is used in telephone dialing; each dialed digit is encoded as two tones sounded simultaneously, sampled at 8000 Hz. Write a script that will receive a series of digits simulating a telephone number to be dialed which will then create an appropriate DTMF audio waveform. This waveform will then be input into a second script which will decode the audio waveform and provide as an output the string of dialed digits. Use the list of frequencies given below. Each portion encoding a given transmit digit should be 0.25 seconds long, and the time between successive digit-encoded waveforms should be 0.125 second. There should be a user-specified amount of white noise added to the signal. Once a transmit waveform has been created, the script should methodically decode the encoded digits using the all-real Goertzel Algorithm that provides a magnitude squared output (see function description below). DFT lengths should be chosen for each DTMF tone so that a particular DTMF frequency being sought will be "on-bin" or nearly so. The dialing digits are encoded according to their column and row, each of which is represented by a particular frequency. The particular dialed digit can be identified by methodically checking the transmitted audio waveform for each of the eight frequencies, thus identifying, for each transmitted digit a particular row and column represented by the two tones in the waveform.

		Col 1	Col 2	Col 3	Col 4
		1209 Hz	1336 Hz	1477 Hz	1633 Hz
Row 1	697 Hz	1	2	3	A
Row 2	770 Hz	4	5	6	B
Row 3	852 Hz	7	8	9	C
Row 4	941 Hz	*	0	#	D

The first step, prior to writing the decoding routine is to discover a useful, relatively short length N of DFT for each DTMF tone. There is no need, for example, to do an 8000-pt Goertzel computation. Instead, use the smallest number of samples that contains an integral number of cycles of the frequency being sought. Thus, the Goertzel Algorithm performed for each DTMF row and column frequency will, in general, be of a different length so that the detection will be "on-bin" and give the best signal to noise ratio. The particular bin being sought obviously changes as well, depending on how many cycles of the tone being sought are contained within the DFT length being performed.

A suggested procedure is as follows:

a. Write a short script that will determine 1) for each DTMF frequency, how many samples per cycle spc there are at the sample rate of 8000 Hz, 2) compute, say, the first 100 multiples of spc, 3) either manually or by code, pick the multiple that is closest to being a whole number N, and note how many cycles of the DTMF frequency this is. The length of DFT to be performed when searching for this DTMF frequency using the Goertzel Algorithm will be N and the bin being sought will be equal to the number of cycles noted.

b. Write a function in the form

function [DFTGBin,MagDFTGBin] = LVxDFTViaGoertzelBin(Bin,Sig)

which returns the DFT Bin value (as *DFTGBin*) and its magnitude when called and supplied with the desired *Bin* (equivalent to frequency k) and a signal *Sig* comprised of N contiguous signal samples where N is the desired DFT length for the particular DTMF frequency being sought determined in step (a) above.

c. Write an encoding function in the format

function OutputWave = LVxEncodeDTMF(DialedDigits,NoiseVar)

where *DialedDigits* is an array of alphanumeric data to be encoded, limited to the alphanumeric data given in the table above, and *NoiseVar* is a number setting the amplitude of white noise to mix with the output audio signal. Let each digit be encoded using 2000 samples of the two tones, sampled at 8000 Hz, followed by 1000 samples of silence. Once all digits have been encoded and the complete audio waveform sequence generated, add white noise to it, weighted by *NoiseVar*.

d. Write a decoding function in the format

function [DialedDigits] = LVxDecodeDTMF(InputSignal)

where *DialedDigits* is the output string of alphanumeric data that was originally encoded, and *InputSignal* is *OutputWave* from your function *LVxEncodeDTMF*. Use the function *LVxDFTVia-GoertzelBin* to search for each row and column frequency in each time period of the input audio waveform that contains an encoded alphanumeric character.

Test your scripts with several call pairs, such as:

(I). **OW = LVxEncodeDTMF([7 0 3],0.05); [DialedDigits] = LVxDecodeDTMF(OW)**

(II) **OW = LVxEncodeDTMF(['A', '0', '3'],0.05); [DialedDigits] = LVxDecodeDTMF(OW)**

Increase the value of *NoiseVar* to see how much noise is necessary to disrupt proper detection. An interesting experiment is to use Goertzel lengths that are constant, such as 200 samples, and adjust the bin value sought to be 200/800 = 1/40 times the DTMF frequency being sought. Using this scheme, you should find that a smaller amount of noise will lead to unreliable detection since the DTMF tones will generally be "off-bin" with this scheme, resulting in a lower signal to noise ratio in general.

16. Use the flow diagram for an 8-pt decimation-in-time FFT (found in Fig. 3.20) to write an expression for Bin 3 of an 8-pt DFT. Determine how many complex multiplications are involved to compute just Bin 3, and compare this to similar complex multiplication counts for a complete 8-pt DFT using direct implementation, a complete 8-pt decimation-in-time FFT, and the Goertzel Algorithm for Bin 3 of an 8-pt DFT. Which is the most economical method to use if it is only necessary to compute a single bin of an 8-pt DFT?

17. Assume that your answer for the previous problem to compute Bin 3 of an 8-pt DFT using the minimum number of operations from the flow diagram is of the form N_{CM} complex multiplica-

tions, how many complex multiplications would be required to compute Bin 3 (or any other bin in particular) of a 1024-pt DFT using the minimum number of operations from the flow diagram?

18. Assume that you have two sequences of length-N to be linearly convolved, and N is a power of 2. Determine 1) how many multiplications are necessary to perform the linear convolution, 2) how many multiplications would be necessary using the DFT method assuming the DFTs are performed by direct implementation, and 3) how many multiplications would be necessary using the DFT method implemented using decimation-in-time FFTs.

19. Use DFTs of the specified lengths to compute samples of the DTFT of the following sequences, and compare results among the different DFT lengths performed on each sequence (comparison should be made using magnitude).

   (a) **[1,zeros(1,29),-1]**; use DFT lengths of 31,32,33,34,64,256,...
   1024,4096
   (b) **[ones(1,39)]**; use DFT lengths of 39,40,41,42,64,256,1024,4096.
   (c) **[real(j.^(0:1:31))]**; use DFT lengths of 32,33,34,35,36,37,38,...
   39,64,256,1024

20. Write the m-code to implement the script

$$LVxInvDFTComputeRect25$$

as described and illustrated in the text.

21. Write a script that will accept as an input a user-specified complex-valued sequence, for which the script will perform the DFT and then synthesize the IDFT harmonic-by-harmonic, displaying the real and imaginary parts of the original sequence, the real and imaginary parts of the cumulative IDFT, which is accumulated harmonic-by-harmonic (press any key for the next harmonic); provide additional subplots to display the real and imaginary parts of both the positive and negative frequencies for each harmonic. The figure and subplots created by the script *LVxInvDFTComputeRect25* as shown in the text should serve as a model (except that for your script the original signal and cumulative IDFT will both need to have real and imaginary subplots). There should be a total of eight subplots on your figure. The script should follow this format, and be tested with the calls shown.

   function LVxInvDFTComplex(InputSignal)
   % InputSignal can be real or complex

   Test your script with the following sequences:

   (a) **LVxInvDFTComplex([ones(1,4),-ones(1,4)])**
   (b) **LVxInvDFTComplex([0, ones(1,3),-ones(1,3)])**
   (c) **LVxInvDFTComplex([1, ones(1,3),-ones(1,3)])**
   (d) **LVxInvDFTComplex([ones(1,8) +j*((-1).^(0:1:7))])**
   (e) **LVxInvDFTComplex([0,j*ones(1,7),0,-j*ones(1,7)])**
   (f) **LVxInvDFTComplex([ones(1,8)] +...**

[0,j*ones(1,3),0,-j*ones(1,3)])

22. Write a function that will receive as arguments 1) a sequence that is either a time domain sequence or the DFT of a time domain sequence, and 2) a flag telling the script to either perform the DFT on the sequence or the IDFT on the sequence. Write the function so that only one DFT algorithm is present, namely a direct-implementation that you write, and that when the flag specifies the IDFT, the proper steps are taken to prepare the sequence and treat the DFT output in accordance with Eq. (3.33) to result in the IDFT of the input sequence. The function should take the form

> **function [OutputSequence] = LVxDFTorIDFT(InputSequence,...**
> **DftOrIdftFLAG)**
> **% InputSequence is either a time domain sequence to have the**
> **% DFT performed on it or a DFT to be turned into a time domain**
> **% sequence via the IDFT, which is performed with the same DFT**
> **% algorithm. Pass DftOrIdftFLAG as 0 to perform the DFT or 1**
> **% to perform the IDFT**
> **% Sample Calls:**
> **% [OutputSeq] = LVxDFTorIDFT([ones(1,8)],0)**
> **% [OutputSeq] = LVxDFTorIDFT([ones(1,8)],1)**
> **% [OutputSeq] = LVxDFTorIDFT([8, zeros(1,7)],0)**
> **% [OutputSeq] = LVxDFTorIDFT([8, zeros(1,7)],1)**

23. Compute the following IDFTs using the matrix method of Eq. (3.31).

> (a) $[10,(-2 + 2j),-2,(-2 - 2j)]$
> (b) $[2,(1 + 1j),0,(1 - 1j)]$
> (c) $[-0.8557,(1.9849 - 2.2580j),(-1.4752 -1.9559j),(-1.4752 +1.9559j),(1.9849 + 2.2580j)]$
> (d) $[1,(-0.5813 - 0.4223j),(0.0813 + 0.2501j),(0.0813 - 0.2501j),(-0.5813 + 0.4223j)]$

24. For the following sequences, sampled over the duration and the rates stated, give the bin width of the resulting DFT, the total number of DFT bins and the total number of unique bins (i.e., for a real sequence, bins representing zero and positive frequencies).

> (a) Sample Duration: 4 seconds; Sample Rate: 4000 Hz.
> (b) Sample Duration: 0.25 seconds; Sample Rate: 8000 Hz.
> (c) Sample Duration: 0.02 seconds; Sample Rate: 44,100 Hz.
> (d) Sample Duration: 15 seconds; Sample Rate: 100 Hz.

25. Write a script that evaluates the frequency-discrimination performance of four different windows in noise as is done by the script *LVxWindowingDisplay*. The script constructs a signal of random noise of standard deviation k and adds to it two sinusoids of unity amplitude having frequencies *freq1* and *freq2*, windows the signal with one of the test windows (*rectwin*, *hamming*, *blackman*, *kaiser* (5)), then obtains the magnitude of the DFT (use DFTs of length 256 or greater). This process is repeated 30 times, and the results averaged. Each window is tested in this manner, and the composite results are plotted and a logarithmic scale. Run one of the sample calls for the script *LVxWindowingDisplay*

to see the results for the four windows. Follow the format below, and test the script with the sample calls given.

function LVxWindowingDisplay(k,freq1,freq2)

% k determines the amount of random noise to be mixed with two unity-amplitude sinusoids having frequencies of freq1 and freq2.

% Test calls are:

% **LVxWindowingDisplay(1,66,68)**

% **LVxWindowingDisplay(1,66.5,68.6)**

% **LVxWindowingDisplay(1,66.5,70.7)**

% **LVxWindowingDisplay(2,66.6,70.5)**

% **LVxWindowingDisplay(2,60,70)**

26. For each of Figs. 3.33, 3.34, and 3.35, verify that the frequency responses shown in plot (d) of each of the figures are correct by creating a cosine signal of the specified length and frequency, windowing it as described for each figure, and then directly obtaining the DTFT, which may be done using a very long DFT on the windowed cosine. Vary the exact placement of the window within the long (say 2^14 samples) cosine signal and note the variation in the DTFT.

27. Write the m-code for the script *LVxPhaseShiftViaDFT* as illustrated and described in the text and below:

function LVxPhaseShiftViaDFT

% Creates a signal having the harmonic spectrum of a square

% wave, but with random phases, then resets all phases to 0

% degrees initially using the DFT/IDFT, resulting in a cusped

% waveform, which is shown in the second subplot; thereafter,

% the phase of each bin is shifted a small amount via DFT every

% time any key is pressed. Eventually, all frequencies in the

% cusped waveform are shifted 90 degrees, and the cusped

% waveform is converted in a square wave. The third subplot

% is always a 90 degree phase-shifted version of the waveform

% in the second plot, and thus starts out as a square wave and is

% gradually converted into a cusped waveform as the cusped

% waveform in the second subplot is converted into

% a square wave

28. Write a short script that verifies or illustrates that the DFT of the product of two time domain sequences is $1/N$ times the circular convolution of the DFTs of each, i.e.,

$$DFT(x_1[n]x_2[n]) = \frac{1}{N}(X_1[k] \circledast X_2[k])$$

29. In this project, we'll write a script that can detect the frequency of an interfering sinusoid mixed with an audio signal. To do this, the audio signal is partitioned into small time windows or frames,

and the DFT of each is obtained so that the spectrum is available over small durations throughout the signal. We'll use a finite signal of five seconds' duration, but this procedure (dividing a signal into short frames) would be mandated, for example, when processing an audio signal (in real time) of indefinite duration. For purposes of illustration, we'll compare the DFT of the entire signal to the spectrogram (i.e., spectrum over time) of the signal created by taking the DFT of many short duration frames of the signal. To enhance the detectability of the interfering sinusoid, we'll also analyze a subset of the frames, those having very low mean bin magnitude, which correspond to the relative periods of silence between uttered words in the audio file *drwatsonSR8K.wav*. We'll use several methods to attempt to detect which bin or bins are indicative of a persistent sinusoid, as might occur, for example, in a public address system or the like due to feedback.

We'll use the script created in this project to identify one or more interfering sinusoids. Later in the series, in the chapters on FIR filtering and IIR filtering, found in Volume III, and LMS adaptive filtering, found in Volume IV, we'll use the script created in this exercise to identify an interfering tone and then filter the signal to remove it. In this project, we'll concentrate on breaking the signal into frames arrayed as columns of a matrix, obtaining the DFT of each frame, and identifying one or more persistent frequencies. We'll also reconstruct the test signal from the DFTs of the frames. A simple modification of this technique (doubling the length of each frame using trailing zeros), will be used when we revisit this script in the filtering chapters of Volume III and will allow us to not only perform the interfering or persistent tone identification analysis, but to then use frequency domain filtering (i.e., convolution via DFT multiplication) to remove the persistent tone, and then construct the filtered test signal from the filter-modified frames. This is a useful technique because real-time systems, for example, must work frame-by-frame, often having no more than a few frames of recent history available for analysis, filtering, and signal reconstruction.

If the purpose is frequency analysis only, nonoverlapping frames can be used. Overlap in framing a signal is used in filtering or compression techniques which result in reconstructed frames that are not perfect reconstructions of the original frames, such as with lossy compression techniques like MP3. When the reconstruction is perfect, nonoverlap of frames works well. When the reconstruction is lossy, 50% overlap of windowed frames is useful because it serves as a crossfading operation (i.e., a smooth transition from one frame to the next, in which the volume of one frame gradually fades while that of the other gradually increases) that hides the discontinuities at frame boundaries, which can cause audible clicks or other artifacts. When using 50% frame overlap, use of a nonrectangular window is essential to perform the crossfading operation. Use of 50% overlap also increases the number of "snapshots" of the frequency content per second, i.e., a finer-grained spectrogram.

We now present the script format, argument description, and test calls, followed by a procedure to write the m-code for it.

```
function [ToneFreq,Fnyq,BnSp] = LVx_DetectContTone(A,...
Freq,RorSS,SzWin,OvrLap,AudSig)
% Mixes a tone of amplitude A and frequency Freq with an
```

% audio file and attempts to identify the frequency of the
% interfering tone, which may have a steady-state amplitude
% A when RorSS is passed as 0, or an amplitude that
% linearly ramps from 0 to A over the length of the
% audio file when RorSS is passed as 1. The audio file
% is 'drwatsonSR8K.wav', 'whoknowsSR8k.wav', or white
% noise, which are selected, respectively, by passing
% AudSig as 1, 2, or 3. SzWin is the size of time window in
% samples into which the test signal (audio file plus tone) is
% partitioned for analysis. In partitioning the test signal into
% time windows or frames, an overlap of 0% or 50% of
% SzWin is performed when OvrLap is passed as 0 and 1,
% respectively. A number of figures are created, including
% a first figure displaying mean bin magnitude of all bins
% for each frame, which serves to identify periods of high
% and low energy in the signal, corresponding to active
% speech and background sound, second and third figures
% that are 3-D spectrogram plots of DFT magnitude versus
% Bin and Frame for the complete signal and a version
% based on only the background or relatively low energy
% frames. Fourth and fifth figures display the normalized
% bin derivatives versus frame for the complete signal and
% the low portion of the signal.
% The output arguments consist of a list of possible
% interfering (relatively steady-state) tones in Hz as T, the
% Nyquist rate as F, and the bin spacing in Hz as B, which
% allows another program to determine if a list of frequencies
% provided as ToneFreq contains adjacent bin frequencies
% and are therefore likely to be members of the same spectral
% component.
% Test calls
% [T,F,B] = LVx_DetectContTone(0.011,100,0,512,1,1)
% [T,F,B] = LVx_DetectContTone(0.008,100,0,512,0,1)
% [T,F,B] = LVx_DetectContTone(0.005,94,0,512,0,1)
% [T,F,B] = LVx_DetectContTone(0.07,150,0,512,1,3)
% [T,F,B] = LVx_DetectContTone(0.011,200,0,512,1,1)
% [T,F,B] = LVx_DetectContTone(0.01,200,0,512,1,2)
% [T,F,B] = LVx_DetectContTone(0.1,200,0,512,1,3)
% [T,F,B] = LVx_DetectContTone(0.01,210,0,512,1,1)

% [T,F,B] = LVx_DetectContTone(0.025,200,1,512,1,1)
% [T,F,B] = LVx_DetectContTone(0.02,200,1,512,1,2)
% [T,F,B] = LVx_DetectContTone(0.1,200,1,512,1,3)
% [T,F,B] = LVx_DetectContTone(0.08,500,0,512,1,3)
% [T,F,B] = LVx_DetectContTone(0.02,500,1,512,1,1)
% [T,F,B] = LVx_DetectContTone(0.018,500,1,512,1,2)
% [T,F,B] = LVx_DetectContTone(0.1,500,1,512,1,3)
% [T,F,B] = LVx_DetectContTone(0.005,1000,0,512,0,1)
% [T,F,B] = LVx_DetectContTone(0.004,1000,0,512,1,1)
% [T,F,B] = LVx_DetectContTone(0.002,3000,0,512,1,1)

Here is a suggested procedure to create the script described above:

(a) Open the audio file '*drwatsonSR8K.wav*', scale it to have maximum magnitude of 1.0, and add a sinusoid of amplitude A and frequency $Freq$, thus forming the test signal $TstSig$.

(b) Form a matrix $TDMat$ by partitioning $TstSig$ into frames of length SR, with overlap of $SR/2$ samples (SR must be even). Each column of the matrix $TDMat$ will be a frame. For example, if the audio file were 32 samples long, and the frame size were 8 samples, then the first column of the matrix will comprise samples 1:8, the second column will be samples 5:12, the third column will be samples 9:16, and so forth, until all 32 samples have been used. If the last column is only partially filled with samples, it is completed with zeros. A good way to proceed is to make this action a function according to the following specification:

function OutMat = LVxVector2FramesInMatrix(Sig,SzWin,...
SampsOvrLap)
% Divides an input signal vector Sig into frames of length
% SzWin, with an amount of overlap in samples equal to
% SampsOvrLap. Each column of OutMat is one frame
% of the input Sig.
% Test calls:
% OutMat = LVxVector2FramesInMatrix([0:1:33],8,4)
% OutMat = LVxVector2FramesInMatrix([0:1:33],8,0)
% OutMat = LVxVector2FramesInMatrix([0:1:33],8,1)
% OutMat = LVxVector2FramesInMatrix([0:1:33],8,2)

(c) If 50% overlap has been used, window each column of $TDMat$ with a suitable window such as hamming, etc. When not using overlap, application of a window results in noticeable amplitude modulation of the output signal reconstructed from the windowed frames, so a nonrectangular window should not be applied to the columns (frames) of $TDMat$ in such a case.

(d) Compute a matrix Fty which is the DFT of $TDMat$, computed using the function fft. If no overlap was used and hence no nonrectangular window applied to the frames of $TDMat$, apply a window as the DFT is obtained (i.e., without permanently applying a window to $TDMat$ as this is not wanted in this case).

Each column of the resulting matrix Fty is the DFT of the corresponding frame (column) in $TDMat$. Resize the column length of Fty to eliminate negative bins, i.e., the new column length should include bins $0:1:SzWin/2$.

(e) Form a matrix $Afty$ which is the absolute value of the elements of Fty, and normalize it by dividing it by the magnitude of the largest element.

(f) Determine or identify the lowest magnitude frames by computing the mean of each column of $Afty$, and computing a histogram of the vector of frame (column) mean values. Use the frames, in order of chronological occurrence, whose magnitudes are found below the lowest histogram threshold level to form a matrix $minFrameMat$. The following code will accomplish this:

```
meanframes = mean(Afty);
[Nhist,Xhist] = hist(meanframes);
histthresh = 1; % lowest histogram bin of the standard ten
% linear divisions of the range of values found in the data
[i,j] = find(meanframes < xhist(histthresh));
minFrameMat = Afty(:,j);
```

(g) Several methods should be used to attempt to identify steady-state amplitude or monotonically increasing amplitude sinusoids. The first method identifies the bin or bins that have magnitudes above a given threshold for the largest number of frames. The threshold can be a certain number of dB, for example, above the mean bin magnitude for the entire matrix. A number of candidate bins or frequencies can be selected using this method. A second method attempts to take advantage of the characteristic of a steady-state sinusoid that the derivative of bin magnitude with respect to frame (i.e., time), or an approximation thereof, the difference between bin magnitudes in adjacent frames, normalized by the mean bin magnitude for the bin over all frames (First Order Difference or FOD), is very low compared to that of bins in which no steady-state sinusoid is present. A number of candidate bins (such as those below the lowest one or two histogram threshold levels, or a certain number of standard deviations below mean) can also be selected using this method. A third method looks for bins that have the largest number of positive derivatives (i.e., FODs), indicating a steadily-rising amplitude sinusoid, which might be found, for example, in a public address system when feedback first arises. All three methods should be performed on both the full frame matrix $Afty$ and the minimum magnitude frame matrix $minFrameMat$. It will generally be found that the estimates derived from $minFrameMat$ are better than those from the full frame matrix.

(h) A voting procedure should be used, requiring that candidates appear on at least two of the three lists generated using the three methods described above, in order to be reported as a steady-state or rapidly rising sinusoid in output variable T. Alternately, certain standards, such as some multiple of standard deviations above or below mean, as the case may be, may also be used.

(i) Attempt to determine the relative degrees of reliability of the methods. Note that steady-state sinusoids are less detectable when they are close to other more transient tones in the audio signal itself. Thus, the detectability of steady-state or rising-amplitude tones in the 100-300 Hz range is more difficult than in other frequency ranges. The methods discussed above can nonetheless identify a steady-state or rising-amplitude sinusoid in the 100-300 Hz range, especially using the minimum energy frame matrix method, which eliminates many of the transient, high-amplitude signal tones.

(j) Play the test signal file through the computer's audio system to determine the relative audibility of the interfering sinusoid throughout the signal. Note that at very low amplitude levels near the threshold of detection using the methods discussed above, the signal, especially when in the low frequency range (say, 100-300 Hz), may be masked at times.

(k) Reconstruct the original test signal from the matrix Fty (this is essentially the reverse of the procedure that created Fty) and play it through the computer's audio system to verify that it is correct. To perform the reconstruction, write a script having the following format:

function tdSig = LVxTDMat2SigVec(TDMat,SzWin,...
SampsOverLap)
% Generates a signal vector tdSig from a matrix of time
% domain frames of a signal, each frame of length SzWin,
% with SampsOverLap samples of overlap. The code
% determines which dimension of TDMat matches SzWin
% and assumes that the other dimension is the frame index.
% If neither dimension of TDMat matches SzWin, an error
% is thrown.
% Test call pair for OverLap=0
% OutMat = LVxVector2FramesInMatrix([0:1:33],8,0)
% tdSig = LVxTDMat2SigVec(OutMat,8,0)
% The following four lines of code, when run in sequence,
% return the input vector [0:1:33]:
% OutMat = LVxVector2FramesInMatrix([0:1:33],8,4);
% szO = size(OutMat);
% OutMat = OutMat.*(triang(8)*ones(1,szO(2)))
% tdSig = LVxTDMat2SigVec(OutMat,8,4)

The call

$$[T,F,B] = LVx\_DetectContTone(0.008,1000,0,512,1,1)$$

results in the following output values (note that the two low frequencies are present in the audio file prior to addition of the 1000 Hz tone):

ToneFreq = [46.9, 62.5, 1000]
Fnyq = 4000
BnSp = 15.625

and the following figures, as described above.

30. Write the m-code for the script

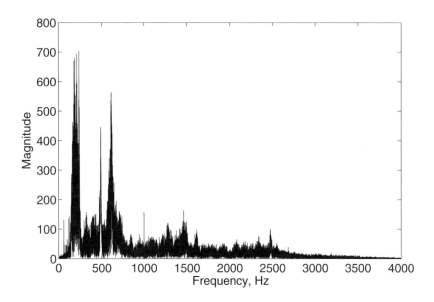

Figure 3.56: DFT (positive frequencies only) of the entire test audio signal, comprised of the audio file 'drwatsonSR8K.wav', normalized to maximum unity magnitude, to which has been added a sine wave of amplitude 0.008 and frequency of 1000 Hz.

```
function LVxZxformFromSamps(b,a,NumzSamps,M,nVals)
% Receives a set of coefficients [b,a] representative of the
% z-transform of a sequence x[n] for which the unit circle is
% included in the ROC. NumzSamps samples of the z-transform
% of x[n] are computed, and then the complete z-transform is
% computed from the samples. From this, nVals samples of x[n]
% are reconstructed using numerical contour integration. The
% samples of X(z) are also used to reconstruct a periodic version
% of x[n] over nVals samples. Both reconstructed sequences, x[n]
% and the periodic version are plotted for comparison. M is the
% number of dense grid z-samples to use for the contour integration
% Test call:
% LVxZxformFromSamps([1,1,1,1],[1],6,5000,10)
% LVxZxformFromSamps([1],[1,0,0.64],36,5000,100)
```

Theoretically, $x[n]$ must be finite length, i.e., identically zero for $n < 0$ and $n \geq N - 1$. Obviously, however, many IIR impulse responses decay away to a magnitude of essentially zero in a finite number of samples, so provided that the number of z-transform samples obtained is

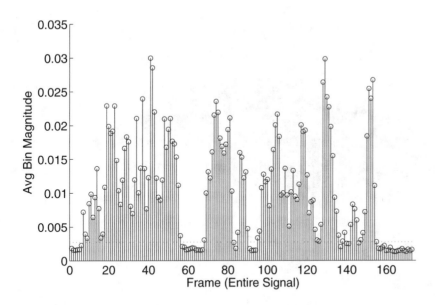

Figure 3.57: A plot of mean bin magnitude versus frame for the entire test signal. Mean bin magnitude is obtained by averaging the bin magnitudes for an entire frame.

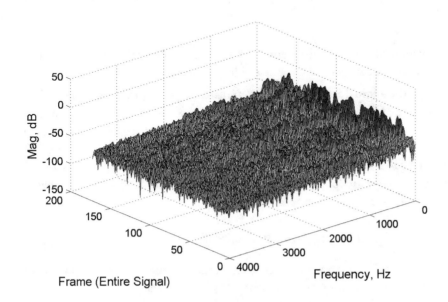

Figure 3.58: Bin magnitude versus Frequency and Frame for the entire test signal. The low-amplitude, steady-state sinusoid at 1000 Hz is barely discernible.

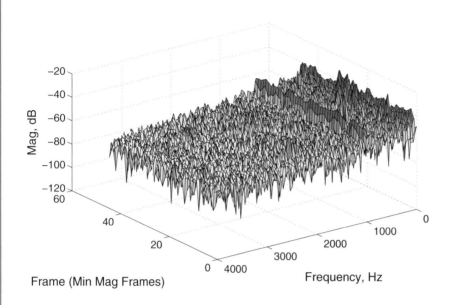

Figure 3.59: Bin magnitude versus Frequency and Frame for the low-magnitude frame matrix (*minFrameMat*). The low-amplitude, steady-state sinusoid at 1000 Hz is readily discernible in the low-energy frames of the signal. Note also an apparent steady-state sinusoid around 60 Hz; this is present in the original audio file.

large enough, reconstruction of $X(z)$ can be performed to a reasonable approximation. This is demonstrated by the second test call given above, the result of which is shown in Fig. 3.62.

31. Evaluate the DTFT of the original sequence, sinc-interpolated sequence, and linear-interpolated sequence created by the script *LVxInterp8Kto11025*, which was developed in Volume I of the series, for the chapter on sampling and binary representation. For the interpolated sequences, evaluate the DTFT both before and after the post-interpolation lowpass filtering. For each evaluation, use 1024 signal samples, perform a DFT of length 2^15, and plot the positive frequency response only in decibels, with the maximum response for each plot being zero dB. Figures 3.63 and 3.64 show the plots that should be created, for example, in response to the call

LVxInterp8Kto11025(0,2950)

32. Write the m-code for the following function specification, the purpose of which is to verify the correctness of Eq. (3.34) in the text. Test your script with the given test calls.

function LVxIDFTviaPosK(TestSig)
% Test signal must be real
% Computes the dft of TestSig, then computes

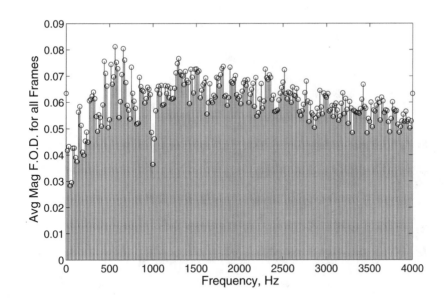

Figure 3.60: A plot of first order differences of bin magnitudes between adjacent frames. The first order differences are all normalized by the respective mean bin magnitudes over all frames in the matrix, which in this case includes all frames of the test signal.

% the IDFT two ways, using formulas (see text)
% for the IDFT that assume bins +k and -k are
% complex conjugates of each other. TestSig is also
% reconstructed via standard IDFT using the
% function ifft, and a figure with three plots is
% generated, the first plot showing TestSig, the
% second one showing the reconstruction via
% the text formula(s), and the third is TestSig
% reconstructed via the function ifft.
% Test calls:
% LVxIDFTviaPosK(ones(1,8))
% LVxIDFTviaPosK(ones(1,9))
% LVxIDFTviaPosK(randn(1,9))
% LVxIDFTviaPosK(j*randn(1,9))

33. Write the m-code for the script

$$LVxDFT Equalization(tstSig, p, k, SR, xplotlim)$$

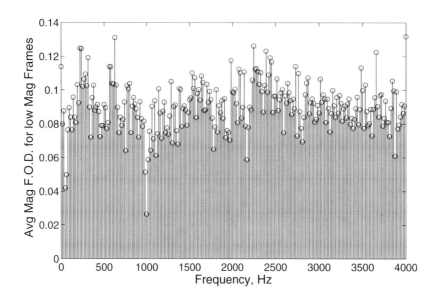

Figure 3.61: A plot of first order differences of bin magnitudes between adjacent frames. Bin indices here are converted to their Hertz equivalents so that a direct search for low FOD can be made by frequency. The first order differences are all normalized by the respective mean bin magnitudes over all frames in the matrix, which in this case includes only low-energy frames of the test signal. Note the large dip in the magnitude of FOD at 500 Hz, indicative of a relatively steady-amplitude sinusoid at that frequency.

as described in the text and according to the following function specification:

function LVxDFTEqualization(tstSig,p,k,SR,xplotlim)
% Creates a test signal; tstSig=1 gives 32 periods of a
% length-16 square wave followed by 16 samples equal to zero;
% tstSig=2 yields 64 periods of a length-16 square wave;
% tstSig = 3 yields a length-1024 chirp from 0 to 512 Hz.
% p is a single real pole to be used to generate a decaying
% magnitude profile which weights SR/4 samples of random noise
% of standard deviation k to generate the test channel
% impulse response. SR is the length of FFT to be used to
% model the channel impulse response and create the inverse
% filter.
% xplotlim is the number of samples of each of the relevant signal
% to plot; the relevant signals are the test signal, the distorted
% test signal, and two equalized versions created, respectively, by
% frequency domain (FD) deconvolution and time domain(TD)

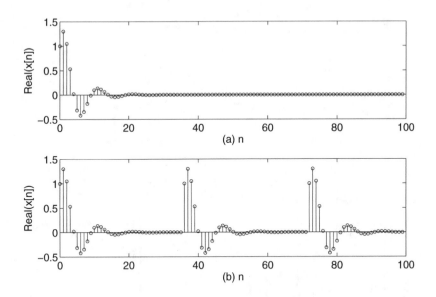

Figure 3.62: (a) A reconstruction of the first 101 samples of the impulse response of the IIR defined by its coefficients $[b, a] = [1],[1,-1.3,0.64]$, obtained by contour integration of $X(z)$, which was computed from 36 samples of $X(z)$; (b) A reconstruction of a periodic version of the same impulse response, computed as the inverse DFS of the 36 samples of $X(z)$, evaluated for $n = 0:1:100$.

% deconvolution.
% Since the impulse response is random noise with a decaying
% magnitude,every call to the function generates a completely
% different impulse response for the simulated channel, and the
% distorted test signal will have a different appearance.
% LVxDFTEqualization(1,0.9,0.5,2048,250)
% LVxDFTEqualization(2,0.9,0.5,2048,250)
% LVxDFTEqualization(3,0.9,0.5,2048,250)

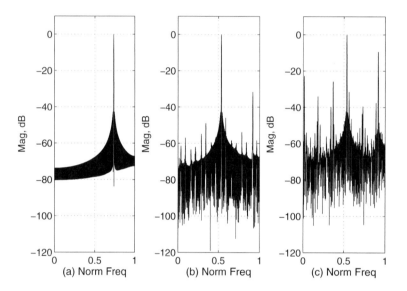

Figure 3.63: (a) DTFT of 1024 samples of original sequence of a 2950 Hz cosine sampled at 8000 Hz; (b) DTFT of 1024 samples of the 2950 Hz cosine after being resampled at 11025 Hz using sinc interpolation, without post-interpolation lowpass filtering; (c) DTFT of 1024 samples of the 2950 Hz cosine after being resampled at 11025 Hz using linear interpolation, without post-interpolation lowpass filtering.

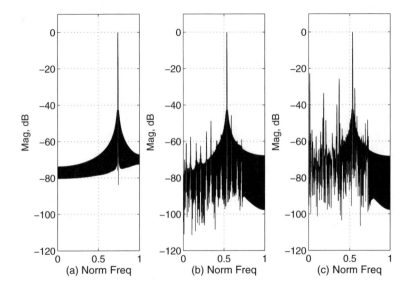

Figure 3.64: (a) DTFT of 1024 samples of original sequence of a 2950 Hz cosine sampled at 8000 Hz; (b) DTFT of 1024 samples of the 2950 Hz cosine after being resampled at 11025 Hz using sinc interpolation, and lowpass filtered with a cutoff frequency of 4 kHz; (c) DTFT of 1024 samples of the 2950 Hz cosine after being resampled at 11025 Hz using linear interpolation, and lowpass filtered with a cutoff frequency of 4 kHz.

APPENDIX A

Software for Use with this Book

A.1 FILE TYPES AND NAMING CONVENTIONS

The text of this book describes many computer programs or scripts to perform computations that illustrate the various signal processing concepts discussed. The computer language used is usually referred to as **m-code** (or as an **m-file** when in file form, using the file extension **.m**) in MATLAB -related literature or discussions, and as **MathScript** in LabVIEW-related discussions (the terms are used interchangeably in this book).

The MATLAB and LabVIEW implementations of m-code (or MathScript) differ slightly (Lab-VIEW's version, for example, at the time of this writing, does not implement Handle Graphics, as does MATLAB).

The book contains mostly scripts that have been tested to run on both MATLAB and Lab-VIEW; these scripts all begin with the letters **LV** and end with the file extension **.m**. Additionally, scripts starting with the letters **LVx** are intended as exercises, in which the student is guided to write the code (the author's solutions, however, are included in the software package and will run when properly called on the Command Line).

Examples are:

LVPlotUnitImpSeq.m

LVxComplexPowerSeries.m

There are also a small number m-files that will run only in MATLAB, as of this writing. They all begin with the letters *ML*. An example is:

ML_SinglePole.m

Additionally, there are a number of LabVIEW Virtual Instruments (VIs) that demonstrate various concepts or properties of signal processing. These have file names that all begin with the letters *Demo* and end with the file extension *.vi*. An example is:

DemoComplexPowerSeriesVI.vi

Finally, there are several sound files that are used with some of the exercises; these are all in the .wav format. An example is:

drwatsonSR4K.wav

A.2   DOWNLOADING THE SOFTWARE

All of the software files needed for use with the book are available for download from the following website:

http://www.morganclaypool.com/page/isen

The entire software package should be stored in a single folder on the user's computer, and the full file name of the folder must be placed on the MATLAB or LabVIEW search path in accordance with the instructions provided by the respective software vendor.

A.3   USING THE SOFTWARE

In MATLAB, once the folder containing the software has been placed on the search path, any script may be run by typing the name (without the file extension, but with any necessary input arguments in parentheses) on the Command Line in the Command Window and pressing *Return*.

In LabVIEW, from the Getting Started window, select MathScript Window from the Tools menu, and the Command Window will be found in the lower left area of the MathScript window. Enter the script name (without the file extension, but with any necessary input arguments in parentheses) in the Command Window and press *Return*. This procedure is essentially the same as that for MATLAB.

Example calls that can be entered on the Command Line and run are

LVAliasing(100,1002)

LV_FFT(8,0)

In the text, many "live" calls (like those just given) are found. All such calls are in boldface as shown in the examples above. When using an electronic version of the book, these can usually be copied and pasted into the Command Line of MATLAB or LabVIEW and run by pressing *Return*. When using a printed copy of the book, it is possible to manually type function calls into the Command Line, but there is also one stored m-file (in the software download package) per chapter that contains clean copies of all the m-code examples from the text of the respective chapter, suitable for copying (these files are described more completely below in the section entitled "Multi-line m-code examples"). There are two general types of m-code examples, single-line function calls and multi-line code examples. Both are discussed immediately below.

A.4   SINGLE-LINE FUNCTION CALLS

The first type of script mentioned above, a named- or defined-function script, is one in which a function is defined; it starts with the word "function" and includes the following, from left to right:

any output arguments, the equal sign, the function name, and, in parentheses immediately following the function name, any input arguments. The function name must always be identical to the file name. An example of a named-function script, is as follows:

function nY = LVMakePeriodicSeq(y,N)
% LVMakePeriodicSeq([1 2 3 4],2)
y = y(:); nY = y*([ones(1,N)]); nY = nY(:)';

For the above function, the output argument is *nY*, the function name is *LVMakePeriodicSeq*, and there are two input arguments, *y* and *N*, that must be supplied with a call to run the function. Functions, in order to be used, must be stored in file form, i.e., as an m-file. The function *LVMakePeriodicSeq* can have only one corresponding file name, which is

LVMakePeriodicSeq.m

In the code above, note that the function definition is on the first line, and an example call that you can paste into the Command Line (after removing or simply not copying the percent sign at the beginning of the line, which marks the line as a comment line) and run by pressing *Return*. Thus you would enter on the Command Line the following, and then press *Return*:

nY = LVMakePeriodicSeq([1,2,3,4],2)

In the above call, note that the output argument has been included; if you do not want the value (or array of values) for the output variable to be displayed in the Command window, place a semicolon after the call:

nY = LVMakePeriodicSeq([1,2,3,4],2);

If you want to see, for example, just the first five values of the output, use the above code to suppress the entire output,and then call for just the number of values that you want to see in the Command window:

nY = LVMakePeriodicSeq([1,2,3,4],2);nY1to5 = nY(1:5)

The result from making the above call is

nY1to5 = [1,2,3,4,1]

A.5   MULTI-LINE M-CODE EXAMPLES

There are also entire multi-line scripts in the text that appear in boldface type; they may or may not include named-functions, but there is always m-code with them in excess of that needed to make a simple function-call. An example might be

N=54; k = 9; x = cos(2*pi*k*(0:1:N-1)/N);
LVFreqResp(x, 500)

Note in the above that there is a named-function (*LVFreqResp*) call, preceded by m-code to define an input argument for the call. Code segments like that above must either be (completely) copied and pasted into the Command Line or manually typed into the Command Line. Copy-and-Paste can often be successfully done directly from a pdf version of the book. This often results in problems (described below), and accordingly, an m-file containing clean copies of most m-code programs from each chapter is supplied with the software package. Most of the calls or multi-line m-code examples from the text that the reader might wish to make are present in m-files such as

McodeVolume1Chapter4.m

McodeVolume2Chapter3.m

and so forth. There is one such file for each chapter of each book, except Chapter 1 of Volume I, which has no m-code examples.

A.6   HOW TO SUCCESSFULLY COPY-AND-PASTE M-CODE

M-code can usually be copied directly from a pdf copy of the book, although a number of minor, easily correctible problems can occur. Two characters, the symbol for raising a number to a power, the circumflex ˆ, and the symbol for vector or matrix transposition, the apostrophe or single quote mark ', are coded for pdf using characters that are non-native to m-code. While these two symbols may look proper in the pdf file, when pasted into the Command line of MATLAB, they will appear in red.

A first way to avoid this copying problem, of course, is simply to use the m-code files described above to copy m-code from. This is probably the most time-efficient method of handling the problem—avoiding it altogether.

A second method to correct the circumflex-and-single-quote problem, if you do want to copy directly from a pdf document, is to simply replace each offending character (circumflex or single quote) by the equivalent one typed from your keyboard. When proper, all such characters will appear in black rather than red in MATLAB. In LabVIEW, the pre-compiler will throw an error at the first such character and cite the line and column number of its location. Simply manually retype/replace each offending character. Since there are usually no more than a few such characters, manually replacing/retyping is quite fast.

Yet a third way (which is usually more time consuming than the two methods described above) to correct the circumflex and apostrophe is to use the function *Reformat*, which is supplied with the software package. To use it, all the copied code from the pdf file is reformatted by hand into one horizontal line, with delimiters (commas or semicolons) inserted (if not already present) where lines have been concatenated. For example, suppose you had copied

```
n = 0:1:4;
y = 2.^n
stem(n,y);
```

where the circumflex is the improper version for use in m-code. We reformat the code into one horizontal line, adding a comma after the second line (a semicolon suppresses computed output on the Command line, while a comma does not), and enclose this string with apostrophes (or single quotes), as shown, where *Reformat* corrects the improper circumflex and *eval* evaluates the string, i.e., runs the code.

$$\text{eval(Reformat('n=0:1:4;y=2.^n;stem(n,y)'))}$$

Occasionally, when copying from the pdf file, essential blank spaces are dropped in the copied result and it is necessary to identify where this has happened and restore the missing space. A common place that this occurs is after a "for" statement. The usual error returned when trying to run the code is that there is an unmatched "end" statement or that there has been an improper use of the reserved word "end". This is caused by the elision of the "for" statement with the ensuing code and is easily corrected by restoring the missing blank space after the "for" statement. Note that the function *Reformat* does not correct for this problem.

A.7 LEARNING TO USE M-CODE

While the intent of this book is to teach the principles of digital signal processing rather than the use of m-code per se, the reader will find that the scripts provided in the text and with the software package will provide many examples of m-code programming starting with simple scripts and functions early in the book to much more involved scripts later in the book, including scripts for use with MATLAB that make extensive use of MATLAB objects such as push buttons, edit boxes, drop-down menus, etc.

Thus the complexity of the m-code examples and exercises progresses throughout the book apace with the complexity of signal processing concepts presented. It is unlikely that the reader or student will find it necessary to separately or explicitly study m-code programming, although it will occasionally be necessary and useful to use the online MATLAB or LabVIEW help files for explanation of the use of, or call syntax of, various built-in functions.

A.8 WHAT YOU NEED WITH MATLAB AND LABVIEW

If you are using a professional edition of MATLAB, you'll need the Signal Processing Toolbox in addition to MATLAB itself. The student version of MATLAB includes the Signal Processing Toolbox.

If you are using either the student or professional edition of LabVIEW, it must be at least Version 8.5 to run the m-files that accompany this book, and to properly run the VIs you'll need the Control Design Toolkit or the newer Control Design and Simulation Module (which is included in the student version of LabVIEW).

APPENDIX B

Vector/Matrix Operations in M-Code

B.1 ROW AND COLUMN VECTORS

Vectors may be either row vectors or column vectors. A typical row vector in m-code might be [3 -1 2 4] or [3,-1,2, 4] (elements in a row can be separated by either commas or spaces), and would appear conventionally as a row:

$$\begin{bmatrix} 3 & -1 & 2 & 4 \end{bmatrix}$$

The same, notated as a column vector, would be [3,-1,2,4]' or [3; -1; 2; 4], where the semicolon sets off different matrix rows:

$$\begin{bmatrix} 3 \\ -1 \\ 2 \\ 4 \end{bmatrix}$$

Notated on paper, a row vector has one row and plural columns, whereas a column vector appears as one column with plural rows.

B.2 VECTOR PRODUCTS

B.2.1 INNER PRODUCT

A row vector and a column vector of the same length as the row vector can be multiplied two different ways, to yield two different results. With the row vector on the left and the column vector on the right,

$$\begin{bmatrix} 1 & 2 & 3 & 4 \end{bmatrix} \begin{bmatrix} 4 \\ 3 \\ 2 \\ 1 \end{bmatrix} = 20$$

corresponding elements of each vector are multiplied, and all products are summed. This is called the **Inner Product**. A typical computation would be

$$[1, 2, 3, 4] * [4; 3; 2; 1] = (1)(4) + (2)(3) + (3)(2) + (4)(1) = 20$$

B.2.2   OUTER PRODUCT

An **Outer Product** results from placing the column vector on the left, and the row vector on the right:

$$
\begin{bmatrix} 4 \\ 3 \\ 2 \\ 1 \end{bmatrix}
\begin{bmatrix} 1 & 2 & 3 & 4 \end{bmatrix}
=
\begin{bmatrix}
4 & 8 & 12 & 16 \\
3 & 6 & 9 & 12 \\
2 & 4 & 6 & 8 \\
1 & 2 & 3 & 4
\end{bmatrix}
$$

The computation is as follows:

$$[4; 3; 2; 1] * [1, 2, 3, 4] = [4, 3, 2, 1; 8, 6, 4, 2; 12, 9, 6, 3; 16, 12, 8, 4]$$

Note that each column in the output matrix is the column of the input column vector, scaled by a column (which is a single value) in the row vector.

B.2.3   PRODUCT OF CORRESPONDING VALUES

Two vectors (or matrices) of exactly the same dimensions may be multiplied on a value-by-value basis by using the notation " .* " (a period followed by an asterisk). Thus two row vectors or two column vectors can be multiplied in this way, and result in a row vector or column vector having the same length as the original two vectors. For example, for two column vectors, we get

$$[1; 2; 3]. * [4; 5; 6] = [4; 10; 18]$$

and for row vectors, we get

$$[1, 2, 3]. * [4, 5, 6] = [4, 10, 18]$$

B.3   MATRIX MULTIPLIED BY A VECTOR OR MATRIX

An m by n matrix, meaning a matrix having m rows and n columns, can be multiplied from the right by an n by 1 column vector, which results in an m by 1 column vector. For example,

$$[1, 2, 1; 2, 1, 2] * [4; 5; 6] = [20; 25]$$

Or, written in standard matrix form:

$$
\begin{bmatrix} 1 & 2 & 1 \\ 2 & 1 & 2 \end{bmatrix}
\begin{bmatrix} 4 \\ 5 \\ 6 \end{bmatrix}
=
\begin{bmatrix} 4 \\ 8 \end{bmatrix}
+
\begin{bmatrix} 10 \\ 5 \end{bmatrix}
+
\begin{bmatrix} 6 \\ 12 \end{bmatrix}
=
\begin{bmatrix} 20 \\ 25 \end{bmatrix}
\tag{B.1}
$$

An m by n matrix can be multiplied from the right by an n by p matrix, resulting in an m by p matrix. Each column of the n by p matrix operates on the m by n matrix as shown in (B.1), and creates another column in the n by p output matrix.

B.4   MATRIX INVERSE AND PSEUDO-INVERSE

Consider the matrix equation

$$\begin{bmatrix} 1 & 4 \\ 3 & -2 \end{bmatrix} \begin{bmatrix} a \\ b \end{bmatrix} = \begin{bmatrix} -2 \\ 3 \end{bmatrix} \tag{B.2}$$

which can be symbolically represented as

$$[M][V] = [C]$$

or simply

$$MV = C$$

and which represents the system of two equations

$$a + 4b = -2$$

$$3a - 2b = 3$$

that can be solved, for example, by scaling the upper equation by -3 and adding to the lower equation

$$-3a - 12b = 6$$

$$3a - 2b = 3$$

which yields

$$-14b = 9$$

or

$$b = -9/14$$

and

$$a = 4/7$$

The inverse of a matrix M is defined as M^{-1} such that

$$MM^{-1} = I$$

where I is called the Identity matrix and consists of all zeros except for the left-to-right downsloping diagonal which is all ones. The Identity matrix is so-called since, for example,

$$\begin{bmatrix} 1 & 0 \\ 0 & 1 \end{bmatrix} \begin{bmatrix} a \\ b \end{bmatrix} = \begin{bmatrix} a \\ b \end{bmatrix}$$

The pseudo-inverse M^{-1} of a matrix M is defined such that

$$M^{-1}M = I$$

System B.2 can also be solved by use of the pseudo-inverse

$$\left[M^{-1} \right] [M][V] = \left[M^{-1} \right] [C]$$

which yields

$$[I][V] = V = \left[M^{-1} \right] [C]$$

In concrete terms, we get

$$\left[M^{-1} \right] \begin{bmatrix} 1 & 4 \\ 3 & -2 \end{bmatrix} \begin{bmatrix} a \\ b \end{bmatrix} = \left[M^{-1} \right] \begin{bmatrix} -2 \\ 3 \end{bmatrix} \tag{B.3}$$

which reduces to

$$\begin{bmatrix} a \\ b \end{bmatrix} = \left[M^{-1} \right] \begin{bmatrix} -2 \\ 3 \end{bmatrix}$$

We can compute the pseudo-inverse M^{-1} and the final solution using the built-in MathScript function *pinv*:

M = [1,4;3,-2];
P = pinv(M)
ans = P*[-2;3]

which yields

$$P = \begin{bmatrix} 0.1429 & 0.2857 \\ 0.2143 & -0.0714 \end{bmatrix}$$

and therefore

$$\begin{bmatrix} a \\ b \end{bmatrix} = \begin{bmatrix} 0.1429 & 0.2857 \\ 0.2143 & -0.0714 \end{bmatrix} \begin{bmatrix} -2 \\ 3 \end{bmatrix}$$

which yields $a = 0.5714$ and $b = -0.6429$ which are the same as 4/7 and -9/14, respectively. A unique solution is possible only when M is square and all rows linearly independent.(a linearly independent row cannot be formed or does not consist solely of a linear combination of other rows in the matrix).

Biography

Forester W. Isen received the B.S. degree from the U. S. Naval Academy in 1971 (majoring in mathematics with additional studies in physics and engineering), and the M. Eng. (EE) degree from the University of Louisville in 1978, and spent a career dealing with intellectual property matters at a government agency working in, and then supervising, the examination and consideration of both technical and legal matters pertaining to the granting of patent rights in the areas of electronic music, horology, and audio and telephony systems (AM and FM stereo, hearing aids, transducer structures, Active Noise Cancellation, PA Systems, Equalizers, Echo Cancellers, etc.). Since retiring from government service at the end of 2004, he worked during 2005 as a consultant in database development, and then subsequently spent several years writing the four-volume series DSP for MATLAB and LabVIEW, calling on his many years of practical experience to create a book on DSP fundamentals that includes not only traditional mathematics and exercises, but "first principle" views and explanations that promote the reader's understanding of the material from an intuitive and practical point of view, as well as a large number of accompanying scripts (to be run on MATLAB or LabVIEW) designed to bring to life the many signal processing concepts discussed in the series.